INTE... CODES

IECC®

A Member of the International Code Family®

INTERNATIONAL
ENERGY CONSERVATION CODE®

2018 International Energy Conservation Code®

First Printing: August 2017
Second Printing: August 2018
Third Printing: March 2019
Fourth Printing: March 2020
Fifth Printing: November 2021

ISBN: 978-1-60983-749-5 (soft-cover edition)

Date of First Publication: August 31, 2017

T029449

PRINTED IN THE USA

PREFACE

Introduction

The *International Energy Conservation Code*® (IECC®) establishes minimum requirements for energy-efficient buildings using prescriptive and performance-related provisions. It is founded on broad-based principles that make possible the use of new materials and new energy-efficient designs. This 2018 edition is fully compatible with all of the *International Codes*® (I-Codes®) published by the International Code Council® (ICC®), including the *International Building Code*®, *International Existing Building Code*®, *International Fire Code*®, *International Fuel Gas Code*®, *International Green Construction Code®*, *International Mechanical Code*®, *International Plumbing Code*®, *International Private Sewage Disposal Code*®, *International Property Maintenance Code*®, *International Residential Code*®, *International Swimming Pool and Spa Code*®, *International Wildland-Urban Interface Code*®, *International Zoning Code*® and *International Code Council Performance Code*®.

This code contains separate provisions for commercial buildings and for low-rise residential buildings (3 stories or less in height above grade). Each set of provisions, IECC—Commercial Provisions and IECC—Residential Provisions, is separately applied to buildings within its respective scope. Each set of provisions is to be treated separately. Each contains a Scope and Administration chapter, a Definitions chapter, a General Requirements chapter, a chapter containing energy efficiency requirements and existing building provisions applicable to buildings within its scope.

The I-Codes, including this *International Energy Conservation Code*, are used in a variety of ways in both the public and private sectors. Most industry professionals are familiar with the I-Codes as the basis of laws and regulations in communities across the U.S. and in other countries. However, the impact of the codes extends well beyond the regulatory arena, as they are used in a variety of nonregulatory settings, including:

- Voluntary compliance programs such as those promoting sustainability, energy efficiency and disaster resistance.
- The insurance industry, to estimate and manage risk, and as a tool in underwriting and rate decisions.
- Certification and credentialing of individuals involved in the fields of building design, construction and safety.
- Certification of building and construction-related products.
- U.S. federal agencies, to guide construction in an array of government-owned properties.
- Facilities management.
- "Best practices" benchmarks for designers and builders, including those who are engaged in projects in jurisdictions that do not have a formal regulatory system or a governmental enforcement mechanism.
- College, university and professional school textbooks and curricula.
- Reference works related to building design and construction.

In addition to the codes themselves, the code development process brings together building professionals on a regular basis. It provides an international forum for discussion and deliberation about building design, construction methods, safety, performance requirements, technological advances and innovative products.

Development

This 2018 edition presents the code as originally issued, with changes reflected in the 2000 through 2015 editions and further changes approved through the ICC Code Development Process through 2017. A new edition such as this is promulgated every 3 years.

This code is founded on principles intended to establish provisions consistent with the scope of an energy conservation code that adequately conserves energy; provisions that do not unnecessarily increase construction costs; provisions that do not restrict the use of new materials, products or methods of construction; and provisions that do not give preferential treatment to particular types or classes of materials, products or methods of construction.

Maintenance

The *International Energy Conservation Code* is kept up to date through the review of proposed changes submitted by code enforcement officials, industry representatives, design professionals and other interested parties. Proposed changes are carefully considered through an open code development process in which all interested and affected parties may participate.

The ICC Code Development Process reflects principles of openness, transparency, balance, due process and consensus, the principles embodied in OMB Circular A-119, which governs the federal government's use of private-sector standards. The ICC process is open to anyone; there is no cost to participate, and people can participate without travel cost through the ICC's cloud-based app, cdpAccess®. A broad cross section of interests are represented in the ICC Code Development Process. The codes, which are updated regularly, include safeguards that allow for emergency action when required for health and safety reasons.

In order to ensure that organizations with a direct and material interest in the codes have a voice in the process, the ICC has developed partnerships with key industry segments that support the ICC's important public safety mission. Some code development committee members were nominated by the following industry partners and approved by the ICC Board:

- National Association of Home Builders (NAHB)
- National Multifamily Housing Council (NMHC)

The code development committees evaluate and make recommendations regarding proposed changes to the codes. Their recommendations are then subject to public comment and council-wide votes. The ICC's governmental members—public safety officials who have no financial or business interest in the outcome—cast the final votes on proposed changes.

The contents of this work are subject to change through the code development cycles and by any governmental entity that enacts the code into law. For more information regarding the code development process, contact the Codes and Standards Development Department of the International Code Council.

While the I-Code development procedure is thorough and comprehensive, the ICC, its members and those participating in the development of the codes disclaim any liability resulting from the publication or use of the I-Codes, or from compliance or noncompliance with their provisions. The ICC does not have the power or authority to police or enforce compliance with the contents of this code.

Code Development Committee Responsibilities (Letter Designations in Front of Section Numbers)

In each code development cycle, proposed changes to the code are considered at the Committee Action Hearings by the applicable International Code Development Committee. The IECC—Commercial Provisions (sections designated with a "C" prior to the section number) are primarily maintained by the Commercial Energy Code Development Committee. The IECC—Residential Provisions (sections designated with an "R" prior to the section number) are maintained by the Residential Energy Code Development Committee. This is designated in the chapter headings by a [CE] and [RE], respectively.

Maintenance responsibilities for the IECC are designated as follows:

[CE] = International Commercial Energy Conservation Code Development Committee

[RE] = International Residential Energy Conservation Code Development Committee

For the development of the 2021 edition of the I-Codes, there will be two groups of code development committees and they will meet in separate years.

Group A Codes (Heard in 2018, Code Change Proposals Deadline: January 8, 2018)	**Group B Codes (Heard in 2019, Code Change Proposals Deadline: January 7, 2019)**
International Building Code – Egress (Chapters 10, 11, Appendix E) – Fire Safety (Chapters 7, 8, 9, 14, 26) – General (Chapters 2–6, 12, 27–33, Appendices A, B, C, D, K, N)	Administrative Provisions (Chapter 1 of all codes except IECC, IRC and IgCC, administrative updates to currently referenced standards, and designated definitions)
International Fire Code	**International Building Code** – Structural (Chapters 15–25, Appendices F, G, H, I, J, L, M)
International Fuel Gas Code	**International Existing Building Code**
International Mechanical Code	**International Energy Conservation Code—Commercial**
International Plumbing Code	**International Energy Conservation Code—Residential** – IECC—Residential – IRC—Energy (Chapter 11)
International Property Maintenance Code	**International Green Construction Code** (Chapter 1)
International Private Sewage Disposal Code	**International Residential Code** – IRC—Building (Chapters 1–10, Appendices E, F, H, J, K, L, M, O, Q, R, S, T)
International Residential Code – IRC—Mechanical (Chapters 12–23) – IRC—Plumbing (Chapters 25–33, Appendices G, I, N, P)	
International Swimming Pool and Spa Code	
International Wildland-Urban Interface Code	
International Zoning Code	
Note: Proposed changes to the ICC *Performance Code*™ will be heard by the code development committee noted in brackets [ ] in the text of the ICC *Performance Code*™.	

Marginal Markings

Solid vertical lines in the margins within the body of the code indicate a technical change from the requirements of the 2015 edition. Deletion indicators in the form of an arrow (➡) are provided in the margin where an entire section, paragraph, exception or table has been deleted or an item in a list of items or a table has been deleted.

Coordination of the International Codes

The coordination of technical provisions is one of the strengths of the ICC family of model codes. The codes can be used as a complete set of complementary documents, which will provide users with full integration and coordination of technical provisions. Individual codes can also be used in subsets or as stand-alone documents. To make sure that each individual code is as complete as possible, some technical provisions that are relevant to more than one subject area are duplicated in some of the model codes. This allows users maximum flexibility in their application of the I-Codes.

Italicized Terms

Selected words and terms defined in Chapter 2, Definitions, are italicized where they appear in code text and the Chapter 2 definition applies. Where such words and terms are not italicized, common-use definitions apply. The words and terms selected have code-specific definitions that the user should read carefully to facilitate better understanding of the code.

Adoption

The International Code Council maintains a copyright in all of its codes and standards. Maintaining copyright allows the ICC to fund its mission through sales of books, in both print and electronic formats. The ICC welcomes adoption of its codes by jurisdictions that recognize and acknowledge the ICC's copyright in the code, and further acknowledge the substantial shared value of the public/private partnership for code development between jurisdictions and the ICC.

The ICC also recognizes the need for jurisdictions to make laws available to the public. All I-Codes and I-Standards, along with the laws of many jurisdictions, are available for free in a nondownloadable form on the ICC's website. Jurisdictions should contact the ICC at adoptions@iccsafe.org to learn how to adopt and distribute laws based on the *International Energy Conservation Code* in a manner that provides necessary access, while maintaining the ICC's copyright.

To facilitate adoption, two sections of this code contain blanks for fill-in information that needs to be supplied by the adopting jurisdiction as part of the adoption legislation. For this code, please see:

Sections C101.1 and R101.1. Insert: [NAME OF JURISDICTION].

EFFECTIVE USE OF THE INTERNATIONAL ENERGY CONSERVATION CODE

The *International Energy Conservation Code* (IECC) is a model code that regulates minimum energy conservation requirements for new buildings. The IECC addresses energy conservation requirements for all aspects of energy uses in both commercial and residential construction, including heating and ventilating, lighting, water heating, and power usage for appliances and building systems.

The IECC is a design document. For example, before one constructs a building, the designer must determine the minimum insulation *R*-values and fenestration *U*-factors for the building exterior envelope. Depending on whether the building is for residential use or for commercial use, the IECC sets forth minimum requirements for exterior envelope insulation, window and door *U*-factors and SHGC ratings, duct insulation, lighting and power efficiency, and water distribution insulation.

Arrangement and Format of the 2018 IECC

The IECC contains two separate sets of provisions—one for commercial buildings and one for residential buildings. Each set of provisions is applied separately to buildings within their scope. The IECC—Commercial Provisions apply to all buildings except for residential buildings three stories or less in height. The IECC—Residential Provisions apply to detached one- and two-family dwellings and multiple single-family dwellings as well as *Group R*-2, R-3 and R-4 buildings three stories or less in height. These scopes are based on the definitions of "Commercial building" and "Residential building," respectively, in Chapter 2 of each set of provisions. Note that the IECC—Commercial Provisions therefore contain provisions for residential buildings four stories or greater in height. Each set of provisions is divided into five different parts:

Chapters	Subjects
1–2	Administration and definitions
3	Climate zones and general materials requirements
4	Energy efficiency requirements
5	Existing buildings
6	Referenced standards

The following is a chapter-by-chapter synopsis of the scope and intent of the provisions of the *International Energy Conservation Code* and applies to both the commercial and residential energy provisions:

Chapter 1 Scope and Administration. This chapter contains provisions for the application, enforcement and administration of subsequent requirements of the code. In addition to establishing the scope of the code, Chapter 1 identifies which buildings and structures come under its purview. Chapter 1 is largely concerned with maintaining "due process of law" in enforcing the energy conservation criteria contained in the body of this code. Only through careful observation of the administrative provisions can the code official reasonably expect to demonstrate that "equal protection under the law" has been provided.

Chapter 2 Definitions. Chapter 2 is the repository of the definitions of terms used in the body of the code. Codes are technical documents and every word, term and punctuation mark can impact the meaning of the code text and the intended results. The code often uses terms that have a unique meaning in the code and the code meaning can differ substantially from the ordinarily understood meaning of the term as used outside of the code.

The terms defined in Chapter 2 are deemed to be of prime importance in establishing the meaning and intent of the code text. The user of the code should be familiar with and consult this chapter because the definitions are essential to the correct interpretation of the code and the user may not be aware that a term is defined.

Additional definitions regarding climate zones are found in Tables 301.3(1) and (2). These are not listed in Chapter 2.

Where understanding of a term's definition is especially key to or necessary for understanding of a particular code provision, the term is shown in *italics*. This is true only for those terms that have a meaning that is unique to the code. In other words, the generally understood meaning of a term or phrase might not be sufficient or consistent with the meaning prescribed by the code; therefore, it is essential that the code-defined meaning be known.

Guidance regarding tense, gender and plurality of defined terms as well as guidance regarding terms not defined in this code is provided.

Chapter 3 General Requirements. Chapter 3 specifies the climate zones that will serve to establish the exterior design conditions. In addition, Chapter 3 provides interior design conditions that are used as a basis for assumptions in heating and cooling load calculations, and provides basic material requirements for insulation materials and fenestration materials.

Climate has a major impact on the energy use of most buildings. The code establishes many requirements such as wall and roof insulation *R*-values, window and door thermal transmittance (*U*-factors) and provisions that affect the mechanical systems based on the climate where the building is located. This chapter contains information that will be used to properly assign the building location into the correct climate zone and is used as the basis for establishing or eliminating requirements.

Chapter 4 Energy Efficiency. Chapter 4 of each set of provisions contains the technical requirements for energy efficiency.

Commercial Energy Efficiency. Chapter 4 of the IECC—Commercial Provisions contains the energy-efficiency-related requirements for the design and construction of most types of commercial buildings and residential buildings greater than three stories in height above grade. This chapter defines requirements for the portions of the building and building systems that impact energy use in new commercial construction and new residential construction greater than three stories in height, and promotes the effective use of energy. In addition to energy conservation requirements for the building envelope, this chapter contains requirements that impact energy efficiency for the HVAC systems, the electrical systems and the plumbing systems. It should be noted, however, that requirements are contained in other codes that have an impact on energy conservation. For instance, requirements for water flow rates are regulated by the *International Plumbing Code*.

Residential Energy Efficiency. Chapter 4 of the IECC—Residential Provisions contains the energy-efficiency-related requirements for the design and construction of residential buildings regulated under this code. It should be noted that the definition of a *residential building* in this code is unique for this code. In this code, a *residential building* is a detached one- and two-family dwelling and multiple single-family dwellings as well as R-2, R-3 or R-4 buildings three stories or less in height. All other buildings, including residential buildings greater than three stories in height, are regulated by the energy conservation requirements in the IECC—Commercial Provisions. The applicable portions of a residential building must comply with the provisions within this chapter for energy efficiency. This chapter defines requirements for the portions of the building and building systems that impact energy use in new residential construction and promotes the effective use of energy. The provisions within the chapter promote energy efficiency in the building envelope, the heating and cooling system and the service water heating system of the building.

Chapter 5 Existing Buildings. Chapter 5 of each set of provisions contains the technical energy efficiency requirements for existing buildings. Chapter 5 provisions address the maintenance of buildings in compliance with the code as well as how additions, alterations, repairs and changes of occupancy need to be addressed from the standpoint of energy efficiency. Specific provisions are provided for historic buildings.

Chapter 6 Referenced Standards. The code contains numerous references to standards that are used to regulate materials and methods of construction. Chapter 6 contains a comprehensive list of all standards that are referenced in the code. The standards are part of the code to the extent of the reference to the standard. Compliance with the referenced standard is necessary for compliance with this code. By providing specifically adopted standards, the construction and installation requirements necessary for compliance with the code can be readily determined. The basis for code compliance is, therefore, established and available on an equal basis to the code official, contractor, designer and owner.

Chapter 6 is organized in a manner that makes it easy to locate specific standards. It lists all of the referenced standards, alphabetically, by acronym of the promulgating agency of the standard. Each agency's standards are then listed in either alphabetical or numeric order based on the standard identification. The list also contains the title of the standard; the edition (date) of the standard referenced; any addenda included as part of the ICC adoption; and the section or sections of this code that reference the standard.

Abbreviations and Notations

The following is a list of common abbreviations and units of measurement used in this code. Some of the abbreviations are for terms defined in Chapter 2. Others are terms used in various tables and text of the code.

AFUE	Annual fuel utilization efficiency
bhp	Brake horsepower (fans)
Btu	British thermal unit
Btu/h-ft^2	Btu per hour per square foot
C-factor	See Chapter 2—Definitions
CDD	Cooling degree days
cfm	Cubic feet per minute
cfm/ft^2	Cubic feet per minute per square foot
ci	Continuous insulation
COP	Coefficient of performance
DCV	Demand control ventilation
°C	Degrees Celsius
°F	Degrees Fahrenheit
DWHR	Drain water heat recovery
DX	Direct expansion
E_c	Combustion efficiency
E_v	Ventilation efficiency
E_t	Thermal efficiency
EER	Energy efficiency ratio
EF	Energy factor
ERI	Energy rating index
F-factor	See Chapter 2—Definitions

FDD	Fault detection and diagnostics
FEG	Fan efficiency grade
FL	Full load
ft^2	Square foot
gpm	Gallons per minute
HDD	Heating degree days
hp	Horsepower
HSPF	Heating seasonal performance factor
HVAC	Heating, ventilating and air conditioning
IEER	Integrated energy efficiency ratio
IPLV	Integrated Part Load Value
Kg/m^2	Kilograms per square meter
kW	Kilowatt
LPD	Light power density (lighting power allowance)
L/s	Liters per second
Ls	Liner system
m^2	Square meters
MERV	Minimum efficiency reporting value
NAECA	National Appliance Energy Conservation Act
NPLV	Nonstandard Part Load Value
Pa	Pascal
PF	Projection factor
pcf	Pounds per cubic foot
psf	Pounds per square foot
PTAC	Packaged terminal air conditioner
PTHP	Packaged terminal heat pump
R-value	See Chapter 2—Definitions
SCOP	Sensible coefficient of performance
SEER	Seasonal energy efficiency ratio
SHGC	Solar Heat Gain Coefficient
SPVAC	Single packaged vertical air conditioner
SPVHP	Single packaged vertical heat pump
SRI	Solar reflectance index
SWHF	Service water heat recovery factor
U-factor	See Chapter 2—Definitions
VAV	Variable air volume
VRF	Variable refrigerant flow
VT	Visible transmittance
W	Watts
w.c.	Water column
w.g.	Water gauge

TABLE OF CONTENTS

IECC—COMMERCIAL PROVISIONS

TABLE OF CONTENTS

CHAPTER 1 [CE]

SCOPE AND ADMINISTRATION

User note:

About this chapter: *Chapter 1 establishes the limits of applicability of the code and describes how the code is to be applied and enforced. Chapter 1 is in two parts: Part 1—Scope and Application and Part 2—Administration and Enforcement. Section 101 identifies what buildings, systems, appliances and equipment fall under its purview and references other I-Codes as applicable. Standards and codes are scoped to the extent referenced.*

The code is intended to be adopted as a legally enforceable document and it cannot be effective without adequate provisions for its administration and enforcement. The provisions of Chapter 1 establish the authority and duties of the code official appointed by the authority having jurisdiction and also establish the rights and privileges of the design professional, contractor and property owner.

PART 1—SCOPE AND APPLICATION

SECTION C101
SCOPE AND GENERAL REQUIREMENTS

C101.1 Title. This code shall be known as the *Energy Conservation Code* of **[NAME OF JURISDICTION]**, and shall be cited as such. It is referred to herein as "this code."

C101.2 Scope. This code applies to *commercial buildings* and the buildings' sites and associated systems and equipment.

C101.3 Intent. This code shall regulate the design and construction of buildings for the effective use and conservation of energy over the useful life of each building. This code is intended to provide flexibility to permit the use of innovative approaches and techniques to achieve this objective. This code is not intended to abridge safety, health or environmental requirements contained in other applicable codes or ordinances.

C101.4 Applicability. Where, in any specific case, different sections of this code specify different materials, methods of construction or other requirements, the most restrictive shall govern. Where there is a conflict between a general requirement and a specific requirement, the specific requirement shall govern.

C101.4.1 Mixed residential and commercial buildings. Where a building includes both *residential building* and *commercial building* portions, each portion shall be separately considered and meet the applicable provisions of IECC—Commercial Provisions or IECC—Residential Provisions.

C101.5 Compliance. *Residential buildings* shall meet the provisions of IECC—Residential Provisions. *Commercial buildings* shall meet the provisions of IECC—Commercial Provisions.

C101.5.1 Compliance materials. The *code official* shall be permitted to approve specific computer software, worksheets, compliance manuals and other similar materials that meet the intent of this code.

SECTION C102
ALTERNATIVE MATERIALS, DESIGN AND METHODS OF CONSTRUCTION AND EQUIPMENT

C102.1 General. The provisions of this code are not intended to prevent the installation of any material or to prohibit any design or method of construction not specifically prescribed by this code, provided that any such alternative has been *approved*. An alternative material, design or method of construction shall be *approved* where the code official finds that the proposed design is satisfactory and complies with the intent of the provisions of this code, and that the material, method or work offered is, for the purpose intended, not less than the equivalent of that prescribed in this code in quality, strength, effectiveness, *fire resistance*, durability and safety. Where the alternative material, design or method of construction is not *approved*, the code official shall respond in writing, stating the reasons why the alternative was not *approved*.

C102.1.1 Above code programs. The *code official* or other authority having jurisdiction shall be permitted to deem a national, state or local energy efficiency program to exceed the energy efficiency required by this code. Buildings *approved* in writing by such an energy efficiency program shall be considered to be in compliance with this code. The requirements identified as "mandatory" in Chapter 4 shall be met.

PART 2—ADMINISTRATION AND ENFORCEMENT

SECTION C103
CONSTRUCTION DOCUMENTS

C103.1 General. Construction documents and other supporting data shall be submitted in one or more sets with each application for a permit. The construction documents shall be prepared by a registered design professional where required by the statutes of the jurisdiction in which the project is to be constructed. Where special conditions exist, the *code official* is authorized to require necessary construction documents to be prepared by a registered design professional.

Exception: The *code official* is authorized to waive the requirements for construction documents or other supporting data if the *code official* determines they are not necessary to confirm compliance with this code.

C103.2 Information on construction documents. Construction documents shall be drawn to scale on suitable material. Electronic media documents are permitted to be submitted where *approved* by the *code official*. Construction documents shall be of sufficient clarity to indicate the location, nature and extent of the work proposed, and show in sufficient detail pertinent data and features of the building, systems and equipment as herein governed. Details shall include, but are not limited to, the following as applicable:

1. Insulation materials and their *R*-values.
2. Fenestration *U*-factors and solar heat gain coefficients (SHGCs).
3. Area-weighted *U*-factor and solar heat gain coefficient (SHGC) calculations.
4. Mechanical system design criteria.
5. Mechanical and service water heating systems and equipment types, sizes and efficiencies.
6. Economizer description.
7. Equipment and system controls.
8. Fan motor horsepower (hp) and controls.
9. Duct sealing, duct and pipe insulation and location.
10. Lighting fixture schedule with wattage and control narrative.
11. Location of *daylight* zones on floor plans.
12. Air sealing details.

C103.2.1 Building thermal envelope depiction. The *building thermal envelope* shall be represented on the construction drawings.

C103.3 Examination of documents. The *code official* shall examine or cause to be examined the accompanying construction documents and shall ascertain whether the construction indicated and described is in accordance with the requirements of this code and other pertinent laws or ordinances. The *code official* is authorized to utilize a registered design professional, or other *approved* entity not affiliated with the building design or construction, in conducting the review of the plans and specifications for compliance with the code.

C103.3.1 Approval of construction documents. When the *code official* issues a permit where construction documents are required, the construction documents shall be endorsed in writing and stamped "Reviewed for Code Compliance." Such *approved* construction documents shall not be changed, modified or altered without authorization from the *code official*. Work shall be done in accordance with the *approved* construction documents.

One set of construction documents so reviewed shall be retained by the *code official*. The other set shall be returned to the applicant, kept at the site of work and shall be open to inspection by the *code official* or a duly authorized representative.

C103.3.2 Previous approvals. This code shall not require changes in the construction documents, construction or designated occupancy of a structure for which a lawful permit has been heretofore issued or otherwise lawfully authorized, and the construction of which has been pursued in good faith within 180 days after the effective date of this code and has not been abandoned.

C103.3.3 Phased approval. The *code official* shall have the authority to issue a permit for the construction of part of an energy conservation system before the construction documents for the entire system have been submitted or *approved*, provided that adequate information and detailed statements have been filed complying with all pertinent requirements of this code. The holders of such permit shall proceed at their own risk without assurance that the permit for the entire energy conservation system will be granted.

C103.4 Amended construction documents. Changes made during construction that are not in compliance with the *approved* construction documents shall be resubmitted for approval as an amended set of construction documents.

C103.5 Retention of construction documents. One set of *approved* construction documents shall be retained by the *code official* for a period of not less than 180 days from date of completion of the permitted work, or as required by state or local laws.

C103.6 Building documentation and closeout submittal requirements. The construction documents shall specify that the documents described in this section be provided to the building owner or owner's authorized agent within 90 days of the date of receipt of the certificate of occupancy.

C103.6.1 Record documents. Construction documents shall be updated to convey a record of the completed work. Such updates shall include mechanical, electrical and control drawings that indicate all changes to size, type and location of components, equipment and assemblies.

C103.6.2 Compliance documentation. Energy code compliance documentation and supporting calculations shall be delivered in one document to the building owner as part of the project record documents or manuals, or as a standalone document. This document shall include the specific energy code edition utilized for compliance determination for each system, documentation demonstrating compliance with Section C303.1.3 for each fenestration product installed, and the interior lighting power compliance path, building area or space-by-space, used to calculate the lighting power allowance.

For projects complying with Item 2 of Section C401.2, the documentation shall include:

1. The envelope insulation compliance path.
2. All compliance calculations including those required by Sections C402.1.5, C403.8.1, C405.3 and C405.4.

For projects complying with Section C407, the documentation shall include that required by Sections C407.4.1 and C407.4.2.

C103.6.3 Systems operation control. Training shall be provided to those responsible for maintaining and operating equipment included in the manuals required by Section C103.6.2.

The training shall include:

1. Review of manuals and permanent certificate.
2. Hands-on demonstration of all normal maintenance procedures, normal operating modes, and all emergency shutdown and startup procedures.
3. Training completion report.

SECTION C104 FEES

C104.1 Fees. A permit shall not be issued until the fees prescribed in Section C104.2 have been paid, nor shall an amendment to a permit be released until the additional fee, if any, has been paid.

C104.2 Schedule of permit fees. A fee for each permit shall be paid as required, in accordance with the schedule as established by the applicable governing authority.

C104.3 Work commencing before permit issuance. Any person who commences any work before obtaining the necessary permits shall be subject to an additional fee established by the *code official* that shall be in addition to the required permit fees.

C104.4 Related fees. The payment of the fee for the construction, *alteration*, removal or demolition of work done in connection to or concurrently with the work or activity authorized by a permit shall not relieve the applicant or holder of the permit from the payment of other fees that are prescribed by law.

C104.5 Refunds. The *code official* is authorized to establish a refund policy.

SECTION C105 INSPECTIONS

C105.1 General. Construction or work for which a permit is required shall be subject to inspection by the code official, his or her designated agent or an approved agency, and such construction or work shall remain visible and able to be accessed for inspection purposes until *approved.* Approval as a result of an inspection shall not be construed to be an approval of a violation of the provisions of this code or of other ordinances of the jurisdiction. Inspections presuming to give authority to violate or cancel the provisions of this code or of other ordinances of the jurisdiction shall not be valid. It shall be the duty of the permit applicant to cause the work to remain visible and able to be accessed for inspection purposes. Neither the code official nor the jurisdiction shall be liable for expense entailed in the removal or replacement of any material, product, system or building component required to allow inspection to validate compliance with this code.

C105.2 Required inspections. The *code official,* his or her designated agent or an approved agency, upon notification, shall make the inspections set forth in Sections C105.2.1 through C105.2.6.

C105.2.1 Footing and foundation insulation. Inspections shall verify the footing and foundation insulation *R*-value, location, thickness, depth of burial and protection of insulation as required by the code, *approved* plans and specifications.

C105.2.2 Thermal envelope. Inspections shall verify the correct type of insulation, *R*-values, location of insulation, fenestration, *U*-factor, SHGC and VT, and that air leakage controls are properly installed, as required by the code, *approved* plans and specifications.

C105.2.3 Plumbing system. Inspections shall verify the type of insulation, *R*-values, protection required, controls and heat traps as required by the code, *approved* plans and specifications.

C105.2.4 Mechanical system. Inspections shall verify the installed HVAC equipment for the correct type and size, controls, insulation, R-values, system and damper air leakage, minimum fan efficiency, energy recovery and economizer as required by the code, *approved* plans and specifications.

C105.2.5 Electrical system. Inspections shall verify lighting system controls, components, and meters as required by the code, *approved* plans and specifications.

C105.2.6 Final inspection. The final inspection shall include verification of the installation and proper operation of all required building controls, and documentation verifying activities associated with required *building commissioning* have been conducted in accordance with Section C408.

C105.3 Reinspection. A building shall be reinspected where determined necessary by the *code official.*

C105.4 Approved inspection agencies. The *code official* is authorized to accept reports of third-party inspection agencies not affiliated with the building design or construction, provided that such agencies are *approved* as to qualifications and reliability relevant to the building components and systems that they are inspecting.

C105.5 Inspection requests. It shall be the duty of the holder of the permit or their duly authorized agent to notify the *code official* when work is ready for inspection. It shall be the duty of the permit holder to provide access to and means for inspections of such work that are required by this code.

C105.6 Reinspection and testing. Where any work or installation does not pass an initial test or inspection, the necessary corrections shall be made to achieve compliance with this code. The work or installation shall then be resubmitted to the *code official* for inspection and testing.

C105.7 Approval. After the prescribed tests and inspections indicate that the work complies in all respects with this code, a notice of approval shall be issued by the *code official.*

C105.7.1 Revocation. The *code official* is authorized to suspend or revoke, in writing, a notice of approval issued under the provisions of this code wherever the certificate is issued in error, or on the basis of incorrect information supplied, or where it is determined that the *building* or structure, premise, or portion thereof is in violation of any ordinance or regulation or any of the provisions of this code.

SECTION C106 VALIDITY

C106.1 General. If a portion of this code is held to be illegal or void, such a decision shall not affect the validity of the remainder of this code.

SECTION C107 REFERENCED STANDARDS

C107.1 Referenced codes and standards. The codes and standards referenced in this code shall be those listed in Chapter 6, and such codes and standards shall be considered as part of the requirements of this code to the prescribed extent of each such reference and as further regulated in Sections C107.1.1 and C107.1.2.

C107.1.1 Conflicts. Where conflicts occur between provisions of this code and referenced codes and standards, the provisions of this code shall apply.

C107.1.2 Provisions in referenced codes and standards. Where the extent of the reference to a referenced code or standard includes subject matter that is within the scope of this code, the provisions of this code, as applicable, shall take precedence over the provisions in the referenced code or standard.

C107.2 Application of references. References to chapter or section numbers, or to provisions not specifically identified by number, shall be construed to refer to such chapter, section or provision of this code.

C107.3 Other laws. The provisions of this code shall not be deemed to nullify any provisions of local, state or federal law.

SECTION C108 STOP WORK ORDER

C108.1 Authority. Where the *code official* finds any work regulated by this code being performed in a manner either contrary to the provisions of this code or dangerous or unsafe, the *code official* is authorized to issue a stop work order.

C108.2 Issuance. The stop work order shall be in writing and shall be given to the owner of the property involved, the owner's authorized agent, or to the person doing the work. Upon issuance of a stop work order, the cited work shall immediately cease. The stop work order shall state the reason for the order and the conditions under which the cited work will be permitted to resume.

C108.3 Emergencies. Where an emergency exists, the *code official* shall not be required to give a written notice prior to stopping the work.

C108.4 Failure to comply. Any person who shall continue any work after having been served with a stop work order, except such work as that person is directed to perform to remove a violation or unsafe condition, shall be liable to a fine as set by the applicable governing authority.

SECTION C109 BOARD OF APPEALS

C109.1 General. In order to hear and decide appeals of orders, decisions or determinations made by the *code official* relative to the application and interpretation of this code, there shall be and is hereby created a board of appeals. The *code official* shall be an ex officio member of said board but shall not have a vote on any matter before the board. The board of appeals shall be appointed by the governing body and shall hold office at its pleasure. The board shall adopt rules of procedure for conducting its business, and shall render all decisions and findings in writing to the appellant with a duplicate copy to the *code official*.

C109.2 Limitations on authority. An application for appeal shall be based on a claim that the true intent of this code or the rules legally adopted thereunder have been incorrectly interpreted, the provisions of this code do not fully apply or an equally good or better form of construction is proposed. The board shall not have authority to waive requirements of this code.

C109.3 Qualifications. The board of appeals shall consist of members who are qualified by experience and training and are not employees of the jurisdiction.

CHAPTER 2 [CE]

DEFINITIONS

User note:

About this chapter: Codes, by their very nature, are technical documents. Every word, term and punctuation mark can add to or change the meaning of a technical requirement. It is necessary to maintain a consensus on the specific meaning of each term contained in the code. Chapter 2 performs this function by stating clearly what specific terms mean for the purposes of the code.

SECTION C201
GENERAL

C201.1 Scope. Unless stated otherwise, the following words and terms in this code shall have the meanings indicated in this chapter.

C201.2 Interchangeability. Words used in the present tense include the future; words in the masculine gender include the feminine and neuter; the singular number includes the plural and the plural includes the singular.

C201.3 Terms defined in other codes. Terms that are not defined in this code but are defined in the *International Building Code, International Fire Code, International Fuel Gas Code, International Mechanical Code, International Plumbing Code* or the *International Residential Code* shall have the meanings ascribed to them in those codes.

C201.4 Terms not defined. Terms not defined by this chapter shall have ordinarily accepted meanings such as the context implies.

SECTION C202
GENERAL DEFINITIONS

ABOVE-GRADE WALL. See "Wall, above-grade."

ACCESS (TO). That which enables a device, appliance or equipment to be reached by ready access or by a means that first requires the removal or movement of a panel, or similar obstruction.

ADDITION. An extension or increase in the *conditioned space* floor area, number of stories or height of a building or structure.

AIR BARRIER. One or more materials joined together in a continuous manner to restrict or prevent the passage of air through the building thermal envelope and its assemblies.

AIR CURTAIN. A device, installed at the building entrance, that generates and discharges a laminar air stream intended to prevent the infiltration of external, unconditioned air into the conditioned spaces, or the loss of interior, conditioned air to the outside.

ALTERATION. Any construction, retrofit or renovation to an existing structure other than repair or addition. Also, a change in a building, electrical, gas, mechanical or plumbing system that involves an extension, addition or change to the arrangement, type or purpose of the original installation.

APPROVED. Acceptable to the code official.

APPROVED AGENCY. An established and recognized agency that is regularly engaged in conducting tests or furnishing inspection services, or furnishing product certification, where such agency has been approved by the *code official*.

AUTOMATIC. Self-acting, operating by its own mechanism when actuated by some impersonal influence, as, for example, a change in current strength, pressure, temperature or mechanical configuration (see "Manual").

BELOW-GRADE WALL. See "Wall, below-grade."

BOILER, MODULATING. A boiler that is capable of more than a single firing rate in response to a varying temperature or heating load.

BOILER SYSTEM. One or more boilers, their piping and controls that work together to supply steam or hot water to heat output devices remote from the boiler.

BUBBLE POINT. The refrigerant liquid saturation temperature at a specified pressure.

BUILDING. Any structure used or intended for supporting or sheltering any use or occupancy, including any mechanical systems, service water heating systems and electric power and lighting systems located on the building site and supporting the building.

BUILDING COMMISSIONING. A process that verifies and documents that the selected building systems have been designed, installed, and function according to the owner's project requirements and construction documents, and to minimum code requirements.

BUILDING ENTRANCE. Any door, set of doors, doorway, or other form of portal that is used to gain access to the building from the outside by the public.

BUILDING SITE. A contiguous area of land that is under the ownership or control of one entity.

BUILDING THERMAL ENVELOPE. The basement walls, exterior walls, floors, ceilings, roofs and any other building element assemblies that enclose conditioned space or provide a boundary between conditioned space and exempt or unconditioned space.

***C*-FACTOR (THERMAL CONDUCTANCE).** The coefficient of heat transmission (surface to surface) through a building component or assembly, equal to the time rate of heat flow per unit area and the unit temperature difference between the warm side and cold side surfaces (Btu/h • ft^2 • °F) [W/(m^2 • K)].

CAPTIVE KEY OVERRIDE. A lighting control that will not release the key that activates the override when the lighting is on.

CAVITY INSULATION. Insulating material located between framing members.

CHANGE OF OCCUPANCY. A change in the use of a building or a portion of a building that results in any of the following:

1. A *change of occupancy* classification.
2. A change from one group to another group within an occupancy classification.
3. Any change in use within a group for which there is a change in the application of the requirements of this code.

CIRCULATING HOT WATER SYSTEM. A specifically designed water distribution system where one or more pumps are operated in the service hot water piping to circulate heated water from the water-heating equipment to the fixture supply and back to the water-heating equipment.

CLIMATE ZONE. A geographical region based on climatic criteria as specified in this code.

CODE OFFICIAL. The officer or other designated authority charged with the administration and enforcement of this code, or a duly authorized representative.

COEFFICENT OF PERFORMANCE (COP) – COOLING. The ratio of the rate of heat removal to the rate of energy input, in consistent units, for a complete refrigerating system or some specific portion of that system under designated operating conditions.

COEFFICIENT OF PERFORMANCE (COP) – HEATING. The ratio of the rate of heat delivered to the rate of energy input, in consistent units, for a complete heat pump system, including the compressor and, if applicable, auxiliary heat, under designated operating conditions.

COMMERCIAL BUILDING. For this code, all buildings that are not included in the definition of "Residential building."

COMPUTER ROOM. A room whose primary function is to house equipment for the processing and storage of electronic data and that has a design electronic data equipment power density of less than 20 watts per square foot (20 watts per 0.092 m^2) of conditioned floor area or a connected design electronic data equipment load of less than 10 kW.

CONDENSING UNIT. A factory-made assembly of refrigeration components designed to compress and liquefy a specific refrigerant. The unit consists of one or more refrigerant compressors, refrigerant condensers (air-cooled, evaporatively cooled, or water-cooled), condenser fans and motors (where used) and factory-supplied accessories.

CONDITIONED FLOOR AREA. The horizontal projection of the floors associated with the *conditioned space*.

CONDITIONED SPACE. An area, room or space that is enclosed within the building thermal envelope and is directly or indirectly heated or cooled. Spaces are indirectly heated or cooled where they communicate through openings with conditioned spaces, where they are separated from conditioned spaces by uninsulated walls, floors or ceilings, or where they contain uninsulated ducts, piping or other sources of heating or cooling.

CONTINUOUS INSULATION (ci). Insulating material that is continuous across all structural members without thermal bridges other than fasteners and service openings. It is installed on the interior or exterior or is integral to any opaque surface of the building envelope.

CRAWL SPACE WALL. The opaque portion of a wall that encloses a crawl space and is partially or totally below grade.

CURTAIN WALL. Fenestration products used to create an external nonload-bearing wall that is designed to separate the exterior and interior environments.

DAYLIGHT RESPONSIVE CONTROL. A device or system that provides automatic control of electric light levels based on the amount of daylight in a space.

DAYLIGHT ZONE. That portion of a building's interior floor area that is illuminated by natural light.

DEMAND CONTROL VENTILATION (DCV). A ventilation system capability that provides for the automatic reduction of outdoor air intake below design rates when the actual occupancy of spaces served by the system is less than design occupancy.

DEMAND RECIRCULATION WATER SYSTEM. A water distribution system having one or more recirculation pumps that pump water from a heated water supply pipe back to the heated water source through a cold water supply pipe.

DUCT. A tube or conduit utilized for conveying air. The air passages of self-contained systems are not to be construed as air ducts.

DUCT SYSTEM. A continuous passageway for the transmission of air that, in addition to ducts, includes duct fittings, dampers, plenums, fans and accessory air-handling equipment and appliances.

DWELLING UNIT. A single unit providing complete independent living facilities for one or more persons, including permanent provisions for living, sleeping, eating, cooking and sanitation.

DYNAMIC GLAZING. Any fenestration product that has the fully reversible ability to change its performance properties, including *U*-factor, solar heat gain coefficient (SHGC), or visible transmittance (VT).

ECONOMIZER, AIR. A duct and damper arrangement and automatic control system that allows a cooling system to supply outside air to reduce or eliminate the need for mechanical cooling during mild or cold weather.

ECONOMIZER, WATER. A system where the supply air of a cooling system is cooled indirectly with water that is itself cooled by heat or mass transfer to the environment without the use of mechanical cooling.

ENCLOSED SPACE. A volume surrounded by solid surfaces such as walls, floors, roofs, and openable devices such as doors and operable windows.

ENERGY ANALYSIS. A method for estimating the annual energy use of the *proposed design* and *standard reference design* based on estimates of energy use.

ENERGY COST. The total estimated annual cost for purchased energy for the building functions regulated by this code, including applicable demand charges.

ENERGY RECOVERY VENTILATION SYSTEM. Systems that employ air-to-air heat exchangers to recover energy from exhaust air for the purpose of preheating, precooling, humidifying or dehumidifying outdoor ventilation air prior to supplying the air to a space, either directly or as part of an HVAC system.

ENERGY SIMULATION TOOL. An *approved* software program or calculation-based methodology that projects the annual energy use of a building.

ENTRANCE DOOR. A vertical fenestration product used for occupant ingress, egress and access in nonresidential buildings, including, but not limited to, exterior entrances utilizing latching hardware and automatic closers and containing over 50 percent glazing specifically designed to withstand heavy-duty usage.

EQUIPMENT ROOM. A space that contains either electrical equipment, mechanical equipment, machinery, water pumps or hydraulic pumps that are a function of the building's services.

EXTERIOR WALL. Walls including both above-grade walls and basement walls.

FAN BRAKE HORSEPOWER (BHP). The horsepower delivered to the fan's shaft. Brake horsepower does not include the mechanical drive losses such as that from belts and gears.

FAN EFFICIENCY GRADE (FEG). A numerical rating identifying the fan's aerodynamic ability to convert shaft power, or impeller power in the case of a direct-driven fan, to air power.

FAN SYSTEM BHP. The sum of the fan brake horsepower of all fans that are required to operate at fan system design conditions to supply air from the heating or cooling source to the *conditioned spaces* and return it to the source or exhaust it to the outdoors.

FAN SYSTEM DESIGN CONDITIONS. Operating conditions that can be expected to occur during normal system operation that result in the highest supply fan airflow rate to conditioned spaces served by the system, other than during air economizer operation.

FAN SYSTEM MOTOR NAMEPLATE HP. The sum of the motor nameplate horsepower of all fans that are required to operate at design conditions to supply air from the heating or cooling source to the *conditioned spaces* and return it to the source or exhaust it to the outdoors.

FENESTRATION. Products classified as either skylights or vertical fenestration.

Skylights. Glass or other transparent or translucent glazing material installed at a slope of less than 60 degrees (1.05 rad) from horizontal, including unit skylights, tubular daylighting devices and glazing materials in solariums, sunrooms, roofs and sloped walls.

Vertical fenestration. Windows that are fixed or operable, opaque doors, glazed doors, glazed block and combination opaque and glazed doors composed of glass or other transparent or translucent glazing materials and installed at a slope of not less than 60 degrees (1.05 rad) from horizontal.

FENESTRATION PRODUCT, FIELD-FABRICATED. A fenestration product whose frame is made at the construction site of standard dimensional lumber or other materials that were not previously cut, or otherwise formed with the specific intention of being used to fabricate a fenestration product or exterior door. Field fabricated does not include site-built fenestration.

FENESTRATION PRODUCT, SITE-BUILT. A fenestration designed to be made up of field-glazed or field-assembled units using specific factory cut or otherwise factory-formed framing and glazing units. Examples of site-built fenestration include storefront systems, curtain walls, and atrium roof systems.

***F*-FACTOR.** The perimeter heat loss factor for slab-on-grade floors (Btu/h • ft • °F) [W/(m • K)].

FLOOR AREA, NET. The actual occupied area not including unoccupied accessory areas such as corridors, stairways, toilet rooms, mechanical rooms and closets.

GENERAL LIGHTING. Lighting that provides a substantially uniform level of illumination throughout an area. General lighting shall not include decorative lighting or lighting that provides a dissimilar level of illumination to serve a specialized application or feature within such area.

GREENHOUSE. A structure or a thermally isolated area of a building that maintains a specialized sunlit environment exclusively used for, and essential to, the cultivation, protection or maintenance of plants.

GROUP R. Buildings or portions of buildings that contain any of the following occupancies as established in the *International Building Code*:

1. *Group R*-1.
2. *Group R*-2 where located more than three stories in height above grade plane.
3. *Group R*-4 where located more than three stories in height above grade plane.

HEAT TRAP. An arrangement of piping and fittings, such as elbows, or a commercially available heat trap that prevents thermosyphoning of hot water during standby periods.

HEATED SLAB. Slab-on-grade construction in which the heating elements, hydronic tubing, or hot air distribution system is in contact with, or placed within or under, the slab.

HIGH SPEED DOOR. A nonswinging door used primarily to facilitate vehicular access or material transportation, with a minimum opening rate of 32 inches (813 mm) per second, a minimum closing rate of 24 inches (610 mm) per second and that includes an automatic-closing device.

HISTORIC BUILDING. Any building or structure that is one or more of the following:

1. Listed, or certified as eligible for listing by the State Historic Preservation Officer or the Keeper of the National Register of Historic Places, in the National Register of Historic Places.
2. Designated as historic under an applicable state or local law.
3. Certified as a contributing resource within a National Register-listed, state-designated or locally designated historic district.

HUMIDISTAT. A regulatory device, actuated by changes in humidity, used for automatic control of relative humidity.

IEC DESIGN H MOTOR. An electric motor that meets all of the following:

1. It is an induction motor designed for use with three-phase power.

2. It contains a cage rotor.
3. It is capable of direct-on-line starting.
4. It has four, six or eight poles.
5. It is rated from 0.4 kW to 1600 kW at a frequency of 60 hertz.

IEC DESIGN N MOTOR. An electric motor that meets all of the following:

1. It is an induction motor designed for use with three-phase power.
2. It contains a cage rotor.
3. It is capable of direct-on-line starting.
4. It has two, four, six or eight poles.
5. It is rated from 0.4 kW to 1600 kW at a frequency of 60 hertz.

INFILTRATION. The uncontrolled inward air leakage into a building caused by the pressure effects of wind or the effect of differences in the indoor and outdoor air density or both.

INTEGRATED PART LOAD VALUE (IPLV). A single-number figure of merit based on part-load EER, COP or kW/ton expressing part-load efficiency for air-conditioning and heat pump equipment on the basis of weighted operation at various load capacities for equipment.

ISOLATION DEVICES. Devices that isolate HVAC zones so that they can be operated independently of one another. *Isolation devices* include separate systems, isolation dampers, and controls providing shutoff at terminal boxes.

LABELED. Equipment, materials or products to which have been affixed a label, seal, symbol or other identifying mark of a nationally recognized testing laboratory, approved agency or other organization concerned with product evaluation that maintains periodic inspection of the production of the labeled items and whose labeling indicates either that the equipment, material or product meets identified standards or has been tested and found suitable for a specified purpose.

LINER SYSTEM (Ls). A system that includes the following:

1. A continuous vapor barrier liner membrane that is installed below the purlins and that is uninterrupted by framing members.
2. An uncompressed, unfaced insulation resting on top of the liner membrane and located between the purlins.

For multilayer installations, the last rated *R*-value of insulation is for unfaced insulation draped over purlins and then compressed when the metal roof panels are attached.

LISTED. Equipment, materials, products or services included in a list published by an organization acceptable to the *code official* and concerned with evaluation of products or services that maintains periodic inspection of production of *listed* equipment or materials or periodic evaluation of services and whose listing states either that the equipment, material, product or service meets identified standards or has been tested and found suitable for a specified purpose.

LOW-SLOPED ROOF. A roof having a slope less than 2 units vertical in 12 units horizontal.

LOW-VOLTAGE DRY-TYPE DISTRIBUTION TRANSFORMER. A transformer that is air-cooled, does not use oil as a coolant, has an input voltage less than or equal to 600 volts and is rated for operation at a frequency of 60 hertz.

LUMINAIRE-LEVEL LIGHTING CONTROLS. A lighting system consisting of one or more luminaires with embedded lighting control logic, occupancy and ambient light sensors, wireless networking capabilities and local override switching capability, where required.

MANUAL. Capable of being operated by personal intervention (see "Automatic").

NAMEPLATE HORSEPOWER. The nominal motor output power rating stamped on the motor nameplate.

NEMA DESIGN A MOTOR. A squirrel-cage motor that meets all of the following:

1. It is designed to withstand full-voltage starting and develop locked-rotor torque as shown in paragraph 12.38.1 of NEMA MG 1.
2. It has pull-up torque not less than the values shown in paragraph 12.40.1 of NEMA MG 1.
3. It has breakdown torque not less than the values shown in paragraph 12.39.1 of NEMA MG 1.
4. It has a locked-rotor current higher than the values shown in paragraph 12.35.1 of NEMA MG 1 for 60 hertz and paragraph 12.35.2 of NEMA MG 1 for 50 hertz.
5. It has a slip at rated load of less than 5 percent for motors with fewer than 10 poles.

NEMA DESIGN B MOTOR. A squirrel-cage motor that meets all of the following:

1. It is designed to withstand full-voltage starting.
2. It develops locked-rotor, breakdown, and pull-up torques adequate for general application as specified in Sections 12.38, 12.39 and 12.40 of NEMA MG1.
3. It draws locked-rotor current not to exceed the values shown in Section 12.35.1 for 60 hertz and Section 12.35.2 for 50 hertz of NEMA MG1.
4. It has a slip at rated load of less than 5 percent for motors with fewer than 10 poles.

NEMA DESIGN C MOTOR. A squirrel-cage motor that meets all of the following:

1. Designed to withstand full-voltage starting and develop locked-rotor torque for high-torque applications up to the values shown in paragraph 12.38.2 of NEMA MG1 (incorporated by reference, see A§431.15).
2. It has pull-up torque not less than the values shown in paragraph 12.40.2 of NEMA MG1.
3. It has breakdown torque not less than the values shown in paragraph 12.39.2 of NEMA MG1.
4. It has a locked-rotor current not to exceed the values shown in paragraph 12.35.1 of NEMA MG1 for 60 hertz and paragraph 12.35.2 for 50 hertz.
5. It has a slip at rated load of less than 5 percent.

NETWORKED GUESTROOM CONTROL SYSTEM. A control system, accessible from the front desk or other central location associated with a *Group R*-1 building, that is capable of identifying the occupancy status of each guestroom according to a timed schedule, and is capable of controlling HVAC in each hotel and motel guestroom separately.

NONSTANDARD PART LOAD VALUE (NPLV). A single-number part-load efficiency figure of merit calculated and referenced to conditions other than IPLV conditions, for units that are not designed to operate at AHRI standard rating conditions.

OCCUPANT SENSOR CONTROL. An automatic control device or system that detects the presence or absence of people within an area and causes lighting, equipment or appliances to be regulated accordingly.

ON-SITE RENEWABLE ENERGY. Energy derived from solar radiation, wind, waves, tides, landfill gas, biogas, biomass or the internal heat of the earth. The energy system providing on-site renewable energy shall be located on the project site.

OPAQUE DOOR. A door that is not less than 50-percent opaque in surface area.

POWERED ROOF/WALL VENTILATORS. A fan consisting of a centrifugal or axial impeller with an integral driver in a weather-resistant housing and with a base designed to fit, usually by means of a curb, over a wall or roof opening.

PROPOSED DESIGN. A description of the proposed building used to estimate annual energy use for determining compliance based on total building performance.

RADIANT HEATING SYSTEM. A heating system that transfers heat to objects and surfaces within a conditioned space, primarily by infrared radiation.

READY ACCESS (TO). That which enables a device, appliance or equipment to be directly reached, without requiring the removal or movement of any panel or similar obstruction.

REFRIGERANT DEW POINT. The refrigerant vapor saturation temperature at a specified pressure.

REFRIGERATED WAREHOUSE COOLER. An enclosed storage space capable of being refrigerated to temperatures above 32°F (0°C), that can be walked into and has a total chilled storage area of not less than 3,000 square feet (279 m^2).

REFRIGERATED WAREHOUSE FREEZER. An enclosed storage space capable of being refrigerated to temperatures at or below 32°F (0°C), that can be walked into and has a total chilled storage area of not less than 3,000 square feet (279 m^2).

REFRIGERATION SYSTEM, LOW TEMPERATURE. Systems for maintaining food product in a frozen state in refrigeration applications.

REFRIGERATION SYSTEM, MEDIUM TEMPERATURE. Systems for maintaining food product above freezing in refrigeration applications.

REGISTERED DESIGN PROFESSIONAL. An individual who is registered or licensed to practice their respective design profession as defined by the statutory requirements of the professional registration laws of the state or jurisdiction in which the project is to be constructed.

REPAIR. The reconstruction or renewal of any part of an existing building for the purpose of its maintenance or to correct damage.

REROOFING. The process of recovering or replacing an existing roof covering. See "Roof recover" and "Roof replacement."

RESIDENTIAL BUILDING. For this code, includes detached one- and two-family dwellings and multiple single-family dwellings (townhouses) and *Group R*-2, R-3 and R-4 buildings three stories or less in height above grade plane.

ROOF ASSEMBLY. A system designed to provide weather protection and resistance to design loads. The system consists of a roof covering and roof deck or a single component serving as both the roof covering and the roof deck. A roof assembly includes the roof covering, underlayment, roof deck, insulation, vapor retarder and interior finish.

ROOF RECOVER. The process of installing an additional roof covering over an existing roof covering without removing the existing roof covering.

ROOF REPAIR. Reconstruction or renewal of any part of an existing roof for the purpose of its maintenance.

ROOF REPLACMENT. The process of removing the existing roof covering, repairing any damaged substrate and installing a new roof covering.

ROOFTOP MONITOR. A raised section of a roof containing vertical fenestration along one or more sides.

***R*-VALUE (THERMAL RESISTANCE).** The inverse of the time rate of heat flow through a body from one of its bounding surfaces to the other surface for a unit temperature difference between the two surfaces, under steady state conditions, per unit area (h • ft^2 • °F/Btu) [(m^2 • K)/W].

SATURATED CONDENSING TEMPERATURE. The saturation temperature corresponding to the measured refrigerant pressure at the condenser inlet for single component and azeotropic refrigerants, and the arithmetic average of the dew point and *bubble point* temperatures corresponding to the refrigerant pressure at the condenser entrance for zeotropic refrigerants.

SERVICE WATER HEATING. Supply of hot water for purposes other than comfort heating.

SLEEPING UNIT. A room or space in which people sleep, that can include permanent provisions for living, eating, and either sanitation or kitchen facilities but not both. Such rooms and spaces that are part of a dwelling unit are not *sleeping units*.

SMALL ELECTRIC MOTOR. A general purpose, alternating current, single speed induction motor.

SOLAR HEAT GAIN COEFFICIENT (SHGC). The ratio of the solar heat gain entering the space through the fenestration assembly to the incident solar radiation. Solar heat gain includes directly transmitted solar heat and absorbed solar radiation, that is then reradiated, conducted or convected into the space.

STANDARD REFERENCE DESIGN. A version of the *proposed design* that meets the minimum requirements of this code and is used to determine the maximum annual energy use requirement for compliance based on total building performance.

STOREFRONT. A system of doors and windows mulled as a composite fenestration structure that has been designed to resist heavy use. *Storefront* systems include, but are not limited to, exterior fenestration systems that span from the floor level or above to the ceiling of the same story on commercial buildings, with or without mulled windows and doors.

THERMOSTAT. An automatic control device used to maintain temperature at a fixed or adjustable setpoint.

TIME SWITCH CONTROL. An automatic control device or system that controls lighting or other loads, including switching off, based on time schedules.

***U*-FACTOR (THERMAL TRANSMITTANCE).** The coefficient of heat transmission (air to air) through a building component or assembly, equal to the time rate of heat flow per unit area and unit temperature difference between the warm side and cold side air films (Btu/h • ft^2 • °F) [W/(m^2 • K)].

VARIABLE REFRIGERANT FLOW SYSTEM. An engineered direct-expansion (DX) refrigerant system that incorporates a common condensing unit, at least one variable-capacity compressor, a distributed refrigerant piping network to multiple indoor fan heating and cooling units each capable of individual zone temperature control, through integral zone temperature control devices and a common communications network. Variable refrigerant flow utilizes three or more steps of control on common interconnecting piping.

VENTILATION. The natural or mechanical process of supplying conditioned or unconditioned air to, or removing such air from, any space.

VENTILATION AIR. That portion of supply air that comes from outside (outdoors) plus any recirculated air that has been treated to maintain the desired quality of air within a designated space.

VISIBLE TRANSMITTANCE [VT]. The ratio of visible light entering the space through the fenestration product assembly to the incident visible light. Visible transmittance includes the effects of glazing material and frame and is expressed as a number between 0 and 1.

VOLTAGE DROP. A decrease in voltage caused by losses in the wiring systems that connect the power source to the load.

WALK-IN COOLER. An enclosed storage space capable of being refrigerated to temperatures above 32°F (0°C) and less than 55°F (12.8°C) that can be walked into, has a ceiling height of not less than 7 feet (2134 mm) and has a total chilled storage area of less than 3,000 square feet (279 m^2).

WALK-IN FREEZER. An enclosed storage space capable of being refrigerated to temperatures at or below 32°F (0°C) that can be walked into, has a ceiling height of not less than 7 feet (2134 mm) and has a total chilled storage area of less than 3,000 square feet (279 m^2).

WALL, ABOVE-GRADE. A wall associated with the *building thermal envelope* that is more than 15 percent above grade and is on the exterior of the building or any wall that is associated with the *building thermal envelope* that is not on the exterior of the building.

WALL, BELOW-GRADE. A wall associated with the basement or first story of the building that is part of the *building thermal envelope*, is not less than 85 percent below grade and is on the exterior of the building.

WATER HEATER. Any heating appliance or equipment that heats potable water and supplies such water to the potable hot water distribution system.

ZONE. A space or group of spaces within a building with heating or cooling requirements that are sufficiently similar so that desired conditions can be maintained throughout using a single controlling device.

CHAPTER 3 [CE]

GENERAL REQUIREMENTS

User note:

About this chapter: *Chapter 3 addresses broadly applicable requirements that would not be at home in other chapters having more specific coverage of subject matter. This chapter establishes climate zone by U.S. counties and also contains product rating, marking and installation requirements for materials such as insulation, windows, doors and siding.*

SECTION C301
CLIMATE ZONES

C301.1 General. *Climate zones* from Figure C301.1 or Table C301.1 shall be used for determining the applicable requirements from Chapter 4. Locations not indicated in Table C301.1 shall be assigned a *climate zone* in accordance with Section C301.3.

C301.2 Warm humid counties. In Table C301.1, warm humid counties are identified by an asterisk.

C301.3 International climate zones. The *climate zone* for any location outside the United States shall be determined by applying Table C301.3(1) and then Table C301.3(2).

C301.4 Tropical climate zone. The tropical *climate zone* shall be defined as:

1. Hawaii, Puerto Rico, Guam, American Samoa, U.S. Virgin Islands, Commonwealth of Northern Mariana Islands; and
2. Islands in the area between the Tropic of Cancer and the Tropic of Capricorn.

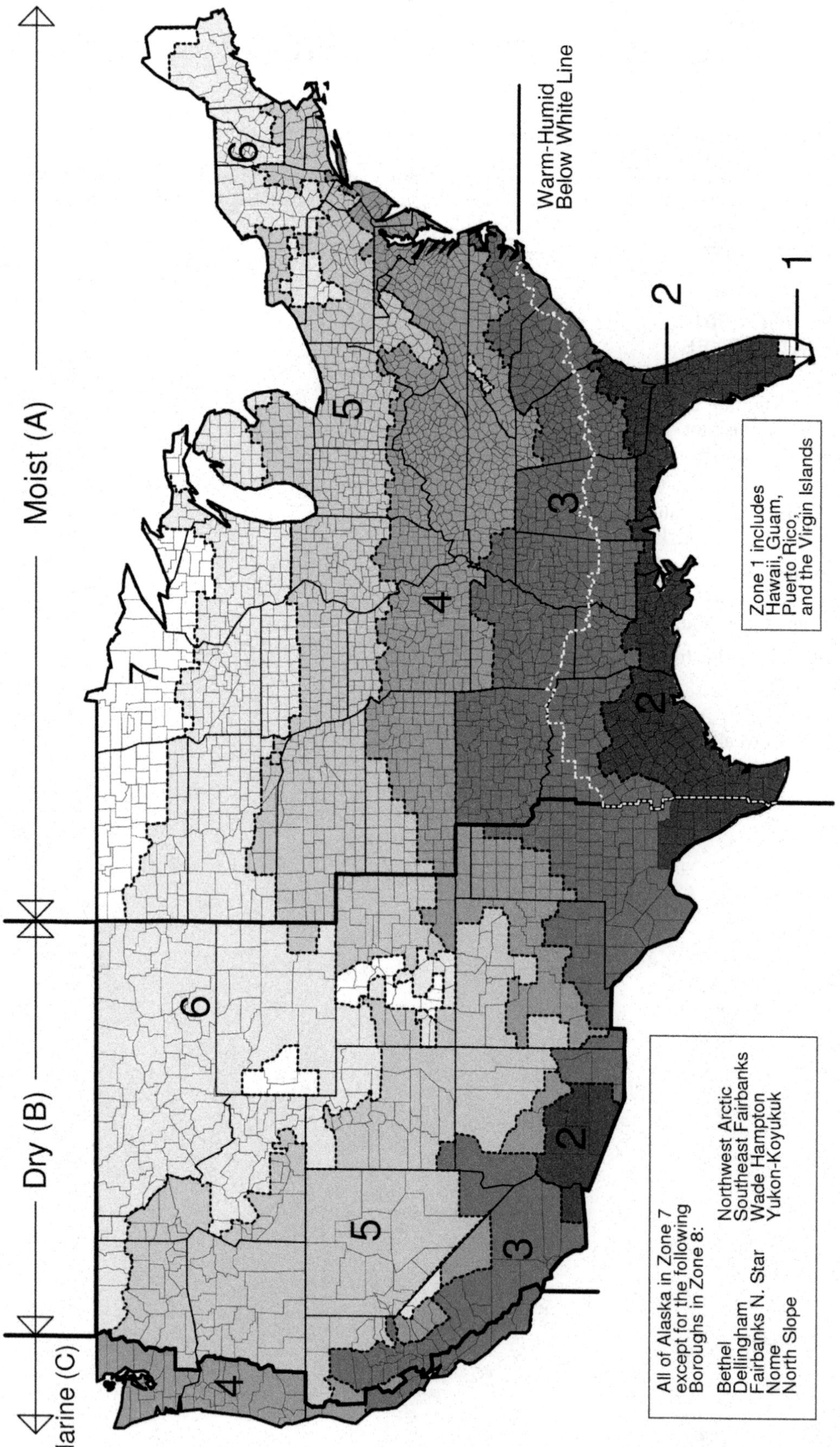

Marine (C)

Dry (B)

Moist (A)

Warm-Humid
Below White Line

7

6

5

4

3

2

1

All of Alaska in Zone 7
except for the following
Boroughs in Zone 8:

Bethel
Dellingham
Fairbanks N. Star
Nome
North Slope
Northwest Arctic
Southeast Fairbanks
Wade Hampton
Yukon-Koyukuk

Zone 1 includes
Hawaii, Guam,
Puerto Rico,
and the Virgin Islands

FIGURE C301.1
CLIMATE ZONES

TABLE C301.1
CLIMATE ZONES, MOISTURE REGIMES, AND WARM-HUMID DESIGNATIONS BY STATE, COUNTY AND TERRITORY

Key: A – Moist, B – Dry, C – Marine. Absence of moisture designation indicates moisture regime is irrelevant. Asterisk (*) indicates a warm-humid location.

US STATES

ALABAMA

3A Autauga*
2A Baldwin*
3A Barbour*
3A Bibb
3A Blount
3A Bullock*
3A Butler*
3A Calhoun
3A Chambers
3A Cherokee
3A Chilton
3A Choctaw*
3A Clarke*
3A Clay
3A Cleburne
3A Coffee*
3A Colbert
3A Conecuh*
3A Coosa
3A Covington*
3A Crenshaw*
3A Cullman
3A Dale*
3A Dallas*
3A DeKalb
3A Elmore*
3A Escambia*
3A Etowah
3A Fayette
3A Franklin
3A Geneva*
3A Greene
3A Hale
3A Henry*
3A Houston*
3A Jackson
3A Jefferson
3A Lamar
3A Lauderdale
3A Lawrence
3A Lee
3A Limestone
3A Lowndes*
3A Macon*
3A Madison
3A Marengo*
3A Marion
3A Marshall
2A Mobile*
3A Monroe*
3A Montgomery*
3A Morgan
3A Perry*
3A Pickens
3A Pike*
3A Randolph
3A Russell*
3A Shelby
3A St. Clair
3A Sumter
3A Talladega
3A Tallapoosa
3A Tuscaloosa
3A Walker
3A Washington*
3A Wilcox*
3A Winston

ALASKA

7 Aleutians East
7 Aleutians West
7 Anchorage
8 Bethel
7 Bristol Bay
7 Denali
8 Dillingham
8 Fairbanks North Star
7 Haines
7 Juneau
7 Kenai Peninsula
7 Ketchikan Gateway
7 Kodiak Island
7 Lake and Peninsula
7 Matanuska-Susitna
8 Nome
8 North Slope
8 Northwest Arctic
7 Prince of Wales-Outer Ketchikan
7 Sitka
7 Skagway-Hoonah-Angoon
8 Southeast Fairbanks
7 Valdez-Cordova
8 Wade Hampton
7 Wrangell-Petersburg
7 Yakutat
8 Yukon-Koyukuk

ARIZONA

5B Apache
3B Cochise
5B Coconino
4B Gila
3B Graham
3B Greenlee
2B La Paz
2B Maricopa
3B Mohave
5B Navajo
2B Pima
2B Pinal
3B Santa Cruz
4B Yavapai
2B Yuma

ARKANSAS

3A Arkansas
3A Ashley
4A Baxter
4A Benton
4A Boone
3A Bradley
3A Calhoun
4A Carroll
3A Chicot
3A Clark
3A Clay
3A Cleburne
3A Cleveland
3A Columbia*
3A Conway
3A Craighead
3A Crawford
3A Crittenden
3A Cross
3A Dallas
3A Desha
3A Drew
3A Faulkner
3A Franklin
4A Fulton
3A Garland
3A Grant
3A Greene
3A Hempstead*
3A Hot Spring
3A Howard
3A Independence
4A Izard
3A Jackson
3A Jefferson
3A Johnson
3A Lafayette*
3A Lawrence
3A Lee
3A Lincoln
3A Little River*
3A Logan
3A Lonoke
4A Madison
4A Marion
3A Miller*
3A Mississippi
3A Monroe
3A Montgomery
3A Nevada
4A Newton
3A Ouachita
3A Perry
3A Phillips
3A Pike
3A Poinsett
3A Polk
3A Pope
3A Prairie
3A Pulaski
3A Randolph
3A Saline
3A Scott
4A Searcy
3A Sebastian
3A Sevier*
3A Sharp
3A St. Francis
4A Stone
3A Union*
3A Van Buren
4A Washington
3A White
3A Woodruff
3A Yell

CALIFORNIA

3C Alameda
6B Alpine
4B Amador
3B Butte
4B Calaveras
3B Colusa
3B Contra Costa
4C Del Norte
4B El Dorado
3B Fresno
3B Glenn

(continued)

**TABLE C301.1—continued
CLIMATE ZONES, MOISTURE REGIMES, AND WARM-HUMID
DESIGNATIONS BY STATE, COUNTY AND TERRITORY**

4C Humboldt
2B Imperial
4B Inyo
3B Kern
3B Kings
4B Lake
5B Lassen
3B Los Angeles
3B Madera
3C Marin
4B Mariposa
3C Mendocino
3B Merced
5B Modoc
6B Mono
3C Monterey
3C Napa
5B Nevada
3B Orange
3B Placer
5B Plumas
3B Riverside
3B Sacramento
3C San Benito
3B San Bernardino
3B San Diego
3C San Francisco
3B San Joaquin
3C San Luis Obispo
3C San Mateo
3C Santa Barbara
3C Santa Clara
3C Santa Cruz
3B Shasta
5B Sierra
5B Siskiyou
3B Solano
3C Sonoma
3B Stanislaus
3B Sutter
3B Tehama
4B Trinity
3B Tulare
4B Tuolumne
3C Ventura
3B Yolo
3B Yuba

COLORADO

5B Adams
6B Alamosa
5B Arapahoe
6B Archuleta
4B Baca
5B Bent
5B Boulder
5B Broomfield
6B Chaffee
5B Cheyenne
7 Clear Creek
6B Conejos
6B Costilla
5B Crowley
6B Custer
5B Delta
5B Denver
6B Dolores
5B Douglas
6B Eagle
5B Elbert
5B El Paso
5B Fremont
5B Garfield
5B Gilpin
7 Grand
7 Gunnison
7 Hinsdale
5B Huerfano
7 Jackson
5B Jefferson
5B Kiowa
5B Kit Carson
7 Lake
5B La Plata
5B Larimer
4B Las Animas
5B Lincoln
5B Logan
5B Mesa
7 Mineral
6B Moffat
5B Montezuma
5B Montrose
5B Morgan
4B Otero
6B Ouray
7 Park
5B Phillips
7 Pitkin
5B Prowers
5B Pueblo
6B Rio Blanco
7 Rio Grande
7 Routt
6B Saguache
7 San Juan
6B San Miguel
5B Sedgwick
7 Summit
5B Teller
5B Washington
5B Weld
5B Yuma

CONNECTICUT

5A (all)

DELAWARE

4A (all)

DISTRICT OF COLUMBIA

4A (all)

FLORIDA

2A Alachua*
2A Baker*
2A Bay*
2A Bradford*
2A Brevard*
1A Broward*
2A Calhoun*
2A Charlotte*
2A Citrus*
2A Clay*
2A Collier*
2A Columbia*
2A DeSoto*
2A Dixie*
2A Duval*
2A Escambia*
2A Flagler*
2A Franklin*
2A Gadsden*
2A Gilchrist*
2A Glades*
2A Gulf*
2A Hamilton*
2A Hardee*
2A Hendry*
2A Hernando*
2A Highlands*
2A Hillsborough*
2A Holmes*
2A Indian River*
2A Jackson*
2A Jefferson*
2A Lafayette*
2A Lake*
2A Lee*
2A Leon*
2A Levy*
2A Liberty*
2A Madison*
2A Manatee*
2A Marion*
2A Martin*
1A Miami-Dade*
1A Monroe*
2A Nassau*
2A Okaloosa*
2A Okeechobee*
2A Orange*
2A Osceola*
2A Palm Beach*
2A Pasco*
2A Pinellas*
2A Polk*
2A Putnam*
2A Santa Rosa*
2A Sarasota*
2A Seminole*
2A St. Johns*
2A St. Lucie*
2A Sumter*
2A Suwannee*
2A Taylor*
2A Union*
2A Volusia*
2A Wakulla*
2A Walton*
2A Washington*

GEORGIA

2A Appling*
2A Atkinson*
2A Bacon*
2A Baker*
3A Baldwin
4A Banks
3A Barrow
3A Bartow
3A Ben Hill*
2A Berrien*
3A Bibb
3A Bleckley*
2A Brantley*
2A Brooks*
2A Bryan*
3A Bulloch*
3A Burke
3A Butts
3A Calhoun*
2A Camden*
3A Candler*
3A Carroll
4A Catoosa
2A Charlton*
2A Chatham*
3A Chattahoochee*
4A Chattooga
3A Cherokee
3A Clarke
3A Clay*
3A Clayton
2A Clinch*
3A Cobb
3A Coffee*
2A Colquitt*
3A Columbia
2A Cook*
3A Coweta

(continued)

**TABLE C301.1—continued
CLIMATE ZONES, MOISTURE REGIMES, AND WARM-HUMID DESIGNATIONS BY STATE, COUNTY AND TERRITORY**

3A Crawford
3A Crisp*
4A Dade
4A Dawson
2A Decatur*
3A DeKalb
3A Dodge*
3A Dooly*
3A Dougherty*
3A Douglas
3A Early*
2A Echols*
2A Effingham*
3A Elbert
3A Emanuel*
2A Evans*
4A Fannin
3A Fayette
4A Floyd
3A Forsyth
4A Franklin
3A Fulton
4A Gilmer
3A Glascock
2A Glynn*
4A Gordon
2A Grady*
3A Greene
3A Gwinnett
4A Habersham
4A Hall
3A Hancock
3A Haralson
3A Harris
3A Hart
3A Heard
3A Henry
3A Houston*
3A Irwin*
3A Jackson
3A Jasper
2A Jeff Davis*
3A Jefferson
3A Jenkins*
3A Johnson*
3A Jones
3A Lamar
2A Lanier*
3A Laurens*
3A Lee*
2A Liberty*
3A Lincoln
2A Long*
2A Lowndes*
4A Lumpkin
3A Macon*
3A Madison
3A Marion*
3A McDuffie
2A McIntosh*
3A Meriwether
2A Miller*
2A Mitchell*
3A Monroe
3A Montgomery*
3A Morgan
4A Murray
3A Muscogee
3A Newton
3A Oconee
3A Oglethorpe
3A Paulding
3A Peach*
4A Pickens
2A Pierce*
3A Pike
3A Polk
3A Pulaski*
3A Putnam
3A Quitman*
4A Rabun
3A Randolph*
3A Richmond
3A Rockdale
3A Schley*
3A Screven*
2A Seminole*
3A Spalding
4A Stephens
3A Stewart*
3A Sumter*
3A Talbot
3A Taliaferro
2A Tattnall*
3A Taylor*
3A Telfair*
3A Terrell*
2A Thomas*
3A Tift*
2A Toombs*
4A Towns
3A Treutlen*
3A Troup
3A Turner*
3A Twiggs*
4A Union
3A Upson
4A Walker
3A Walton
2A Ware*
3A Warren
3A Washington
2A Wayne*
3A Webster*
3A Wheeler*
4A White
4A Whitfield
3A Wilcox*
3A Wilkes
3A Wilkinson
3A Worth*

HAWAII

1A (all)*

IDAHO

5B Ada
6B Adams
6B Bannock
6B Bear Lake
5B Benewah
6B Bingham
6B Blaine
6B Boise
6B Bonner
6B Bonneville
6B Boundary
6B Butte
6B Camas
5B Canyon
6B Caribou
5B Cassia
6B Clark
5B Clearwater
6B Custer
5B Elmore
6B Franklin
6B Fremont
5B Gem
5B Gooding
5B Idaho
6B Jefferson
5B Jerome
5B Kootenai
5B Latah
6B Lemhi
5B Lewis
5B Lincoln
6B Madison
5B Minidoka
5B Nez Perce
6B Oneida
5B Owyhee
5B Payette
5B Power
5B Shoshone
6B Teton
5B Twin Falls
6B Valley
5B Washington

ILLINOIS

5A Adams
4A Alexander
4A Bond
5A Boone
5A Brown
5A Bureau
5A Calhoun
5A Carroll
5A Cass
5A Champaign
4A Christian
5A Clark
4A Clay
4A Clinton
5A Coles
5A Cook
4A Crawford
5A Cumberland
5A DeKalb
5A De Witt
5A Douglas
5A DuPage
5A Edgar
4A Edwards
4A Effingham
4A Fayette
5A Ford
4A Franklin
5A Fulton
4A Gallatin
5A Greene
5A Grundy
4A Hamilton
5A Hancock
4A Hardin
5A Henderson
5A Henry
5A Iroquois
4A Jackson
4A Jasper
4A Jefferson
5A Jersey
5A Jo Daviess
4A Johnson
5A Kane
5A Kankakee
5A Kendall
5A Knox
5A Lake
5A La Salle
4A Lawrence
5A Lee
5A Livingston
5A Logan
5A Macon
4A Macoupin
4A Madison
4A Marion
5A Marshall
5A Mason
4A Massac
5A McDonough
5A McHenry

(continued)

TABLE C301.1—continued
CLIMATE ZONES, MOISTURE REGIMES, AND WARM-HUMID DESIGNATIONS BY STATE, COUNTY AND TERRITORY

5A McLean
5A Menard
5A Mercer
4A Monroe
4A Montgomery
5A Morgan
5A Moultrie
5A Ogle
5A Peoria
4A Perry
5A Piatt
5A Pike
4A Pope
4A Pulaski
5A Putnam
4A Randolph
4A Richland
5A Rock Island
4A Saline
5A Sangamon
5A Schuyler
5A Scott
4A Shelby
5A Stark
4A St. Clair
5A Stephenson
5A Tazewell
4A Union
5A Vermilion
4A Wabash
5A Warren
4A Washington
4A Wayne
4A White
5A Whiteside
5A Will
4A Williamson
5A Winnebago
5A Woodford

INDIANA

5A Adams
5A Allen
5A Bartholomew
5A Benton
5A Blackford
5A Boone
4A Brown
5A Carroll
5A Cass
4A Clark
5A Clay
5A Clinton
4A Crawford
4A Daviess
4A Dearborn
5A Decatur
5A De Kalb
5A Delaware
4A Dubois
5A Elkhart
5A Fayette
4A Floyd
5A Fountain
5A Franklin
5A Fulton
4A Gibson
5A Grant
4A Greene
5A Hamilton
5A Hancock
4A Harrison
5A Hendricks
5A Henry
5A Howard
5A Huntington
4A Jackson
5A Jasper
5A Jay
4A Jefferson
4A Jennings
5A Johnson
4A Knox
5A Kosciusko
5A LaGrange
5A Lake
5A LaPorte
4A Lawrence
5A Madison
5A Marion
5A Marshall
4A Martin
5A Miami
4A Monroe
5A Montgomery
5A Morgan
5A Newton
5A Noble
4A Ohio
4A Orange
5A Owen
5A Parke
4A Perry
4A Pike
5A Porter
4A Posey
5A Pulaski
5A Putnam
5A Randolph
4A Ripley
5A Rush
4A Scott
5A Shelby
4A Spencer
5A Starke
5A Steuben
5A St. Joseph
4A Sullivan
4A Switzerland
5A Tippecanoe
5A Tipton
5A Union
4A Vanderburgh
5A Vermillion
5A Vigo
5A Wabash
5A Warren
4A Warrick
4A Washington
5A Wayne
5A Wells
5A White
5A Whitley

IOWA

5A Adair
5A Adams
6A Allamakee
5A Appanoose
5A Audubon
5A Benton
6A Black Hawk
5A Boone
6A Bremer
6A Buchanan
6A Buena Vista
6A Butler
6A Calhoun
5A Carroll
5A Cass
5A Cedar
6A Cerro Gordo
6A Cherokee
6A Chickasaw
5A Clarke
6A Clay
6A Clayton
5A Clinton
5A Crawford
5A Dallas
5A Davis
5A Decatur
6A Delaware
5A Des Moines
6A Dickinson
5A Dubuque
6A Emmet
6A Fayette
6A Floyd
6A Franklin
5A Fremont
5A Greene
6A Grundy
5A Guthrie
6A Hamilton
6A Hancock
6A Hardin
5A Harrison
5A Henry
6A Howard
6A Humboldt
6A Ida
5A Iowa
5A Jackson
5A Jasper
5A Jefferson
5A Johnson
5A Jones
5A Keokuk
6A Kossuth
5A Lee
5A Linn
5A Louisa
5A Lucas
6A Lyon
5A Madison
5A Mahaska
5A Marion
5A Marshall
5A Mills
6A Mitchell
5A Monona
5A Monroe
5A Montgomery
5A Muscatine
6A O'Brien
6A Osceola
5A Page
6A Palo Alto
6A Plymouth
6A Pocahontas
5A Polk
5A Pottawattamie
5A Poweshiek
5A Ringgold
6A Sac
5A Scott
5A Shelby
6A Sioux
5A Story
5A Tama
5A Taylor
5A Union
5A Van Buren
5A Wapello
5A Warren
5A Washington
5A Wayne
6A Webster
6A Winnebago

(continued)

TABLE C301.1—continued
CLIMATE ZONES, MOISTURE REGIMES, AND WARM-HUMID DESIGNATIONS BY STATE, COUNTY AND TERRITORY

6A Winneshiek
5A Woodbury
6A Worth
6A Wright

KANSAS

4A Allen
4A Anderson
4A Atchison
4A Barber
4A Barton
4A Bourbon
4A Brown
4A Butler
4A Chase
4A Chautauqua
4A Cherokee
5A Cheyenne
4A Clark
4A Clay
5A Cloud
4A Coffey
4A Comanche
4A Cowley
4A Crawford
5A Decatur
4A Dickinson
4A Doniphan
4A Douglas
4A Edwards
4A Elk
5A Ellis
4A Ellsworth
4A Finney
4A Ford
4A Franklin
4A Geary
5A Gove
5A Graham
4A Grant
4A Gray
5A Greeley
4A Greenwood
5A Hamilton
4A Harper
4A Harvey
4A Haskell
4A Hodgeman
4A Jackson
4A Jefferson
5A Jewell
4A Johnson
4A Kearny
4A Kingman
4A Kiowa
4A Labette
5A Lane
4A Leavenworth
4A Lincoln
4A Linn
5A Logan
4A Lyon
4A Marion
4A Marshall
4A McPherson
4A Meade
4A Miami
5A Mitchell
4A Montgomery
4A Morris
4A Morton
4A Nemaha
4A Neosho
5A Ness
5A Norton
4A Osage
5A Osborne
4A Ottawa
4A Pawnee
5A Phillips
4A Pottawatomie
4A Pratt
5A Rawlins
4A Reno
5A Republic
4A Rice
4A Riley
5A Rooks
4A Rush
4A Russell
4A Saline
5A Scott
4A Sedgwick
4A Seward
4A Shawnee
5A Sheridan
5A Sherman
5A Smith
4A Stafford
4A Stanton
4A Stevens
4A Sumner
5A Thomas
5A Trego
4A Wabaunsee
5A Wallace
4A Washington
5A Wichita
4A Wilson
4A Woodson
4A Wyandotte

KENTUCKY

4A (all)

LOUISIANA

2A Acadia*
2A Allen*
2A Ascension*
2A Assumption*
2A Avoyelles*
2A Beauregard*
3A Bienville*
3A Bossier*
3A Caddo*
2A Calcasieu*
3A Caldwell*
2A Cameron*
3A Catahoula*
3A Claiborne*
3A Concordia*
3A De Soto*
2A East Baton Rouge*
3A East Carroll
2A East Feliciana*
2A Evangeline*
3A Franklin*
3A Grant*
2A Iberia*
2A Iberville*
3A Jackson*
2A Jefferson*
2A Jefferson Davis*
2A Lafayette*
2A Lafourche*
3A La Salle*
3A Lincoln*
2A Livingston*
3A Madison*
3A Morehouse
3A Natchitoches*
2A Orleans*
3A Ouachita*
2A Plaquemines*
2A Pointe Coupee*
2A Rapides*
3A Red River*
3A Richland*
3A Sabine*
2A St. Bernard*
2A St. Charles*
2A St. Helena*
2A St. James*
2A St. John the Baptist*
2A St. Landry*
2A St. Martin*
2A St. Mary*
2A St. Tammany*
2A Tangipahoa*
3A Tensas*
2A Terrebonne*
3A Union*
2A Vermilion*
3A Vernon*
2A Washington*
3A Webster*
2A West Baton Rouge*
3A West Carroll
2A West Feliciana*
3A Winn*

MAINE

6A Androscoggin
7 Aroostook
6A Cumberland
6A Franklin
6A Hancock
6A Kennebec
6A Knox
6A Lincoln
6A Oxford
6A Penobscot
6A Piscataquis
6A Sagadahoc
6A Somerset
6A Waldo
6A Washington
6A York

MARYLAND

4A Allegany
4A Anne Arundel
4A Baltimore
4A Baltimore (city)
4A Calvert
4A Caroline
4A Carroll
4A Cecil
4A Charles
4A Dorchester
4A Frederick
5A Garrett
4A Harford
4A Howard
4A Kent
4A Montgomery
4A Prince George's
4A Queen Anne's
4A Somerset
4A St. Mary's
4A Talbot
4A Washington
4A Wicomico
4A Worcester

MASSACHSETTS

5A (all)

MICHIGAN

6A Alcona
6A Alger

(continued)

TABLE C301.1—continued
CLIMATE ZONES, MOISTURE REGIMES, AND WARM-HUMID DESIGNATIONS BY STATE, COUNTY AND TERRITORY

5A Allegan
6A Alpena
6A Antrim
6A Arenac
7 Baraga
5A Barry
5A Bay
6A Benzie
5A Berrien
5A Branch
5A Calhoun
5A Cass
6A Charlevoix
6A Cheboygan
7 Chippewa
6A Clare
5A Clinton
6A Crawford
6A Delta
6A Dickinson
5A Eaton
6A Emmet
5A Genesee
6A Gladwin
7 Gogebic
6A Grand Traverse
5A Gratiot
5A Hillsdale
7 Houghton
6A Huron
5A Ingham
5A Ionia
6A Iosco
7 Iron
6A Isabella
5A Jackson
5A Kalamazoo
6A Kalkaska
5A Kent
7 Keweenaw
6A Lake
5A Lapeer
6A Leelanau
5A Lenawee
5A Livingston
7 Luce
7 Mackinac
5A Macomb
6A Manistee
6A Marquette
6A Mason
6A Mecosta
6A Menominee
5A Midland
6A Missaukee
5A Monroe
5A Montcalm
6A Montmorency
5A Muskegon
6A Newaygo
5A Oakland
6A Oceana
6A Ogemaw
7 Ontonagon
6A Osceola
6A Oscoda
6A Otsego
5A Ottawa
6A Presque Isle
6A Roscommon
5A Saginaw
6A Sanilac
7 Schoolcraft
5A Shiawassee
5A St. Clair
5A St. Joseph
5A Tuscola
5A Van Buren
5A Washtenaw
5A Wayne
6A Wexford

MINNESOTA

7 Aitkin
6A Anoka
7 Becker
7 Beltrami
6A Benton
6A Big Stone
6A Blue Earth
6A Brown
7 Carlton
6A Carver
7 Cass
6A Chippewa
6A Chisago
7 Clay
7 Clearwater
7 Cook
6A Cottonwood
7 Crow Wing
6A Dakota
6A Dodge
6A Douglas
6A Faribault
6A Fillmore
6A Freeborn
6A Goodhue
7 Grant
6A Hennepin
6A Houston
7 Hubbard
6A Isanti
7 Itasca
6A Jackson
7 Kanabec
6A Kandiyohi
7 Kittson
7 Koochiching
6A Lac qui Parle
7 Lake
7 Lake of the Woods
6A Le Sueur
6A Lincoln
6A Lyon
7 Mahnomen
7 Marshall
6A Martin
6A McLeod
6A Meeker
7 Mille Lacs
6A Morrison
6A Mower
6A Murray
6A Nicollet
6A Nobles
7 Norman
6A Olmsted
7 Otter Tail
7 Pennington
7 Pine
6A Pipestone
7 Polk
6A Pope
6A Ramsey
7 Red Lake
6A Redwood
6A Renville
6A Rice
6A Rock
7 Roseau
6A Scott
6A Sherburne
6A Sibley
6A Stearns
6A Steele
6A Stevens
7 St. Louis
6A Swift
6A Todd
6A Traverse
6A Wabasha
7 Wadena
6A Waseca
6A Washington
6A Watonwan
7 Wilkin
6A Winona
6A Wright
6A Yellow Medicine

MISSISSIPPI

3A Adams*
3A Alcorn
3A Amite*
3A Attala
3A Benton
3A Bolivar
3A Calhoun
3A Carroll
3A Chickasaw
3A Choctaw
3A Claiborne*
3A Clarke
3A Clay
3A Coahoma
3A Copiah*
3A Covington*
3A DeSoto
3A Forrest*
3A Franklin*
3A George*
3A Greene*
3A Grenada
2A Hancock*
2A Harrison*
3A Hinds*
3A Holmes
3A Humphreys
3A Issaquena
3A Itawamba
2A Jackson*
3A Jasper
3A Jefferson*
3A Jefferson Davis*
3A Jones*
3A Kemper
3A Lafayette
3A Lamar*
3A Lauderdale
3A Lawrence*
3A Leake
3A Lee
3A Leflore
3A Lincoln*
3A Lowndes
3A Madison
3A Marion*
3A Marshall
3A Monroe
3A Montgomery
3A Neshoba
3A Newton
3A Noxubee
3A Oktibbeha
3A Panola
2A Pearl River*
3A Perry*
3A Pike*

(continued)

TABLE C301.1—continued
CLIMATE ZONES, MOISTURE REGIMES, AND WARM-HUMID DESIGNATIONS BY STATE, COUNTY AND TERRITORY

3A Pontotoc
3A Prentiss
3A Quitman
3A Rankin*
3A Scott
3A Sharkey
3A Simpson*
3A Smith*
2A Stone*
3A Sunflower
3A Tallahatchie
3A Tate
3A Tippah
3A Tishomingo
3A Tunica
3A Union
3A Walthall*
3A Warren*
3A Washington
3A Wayne*
3A Webster
3A Wilkinson*
3A Winston
3A Yalobusha
3A Yazoo

MISSOURI

5A Adair
5A Andrew
5A Atchison
4A Audrain
4A Barry
4A Barton
4A Bates
4A Benton
4A Bollinger
4A Boone
5A Buchanan
4A Butler
5A Caldwell
4A Callaway
4A Camden
4A Cape Girardeau
4A Carroll
4A Carter
4A Cass
4A Cedar
5A Chariton
4A Christian
5A Clark
4A Clay
5A Clinton
4A Cole
4A Cooper
4A Crawford
4A Dade
4A Dallas
5A Daviess
5A DeKalb
4A Dent
4A Douglas
4A Dunklin
4A Franklin
4A Gasconade
5A Gentry
4A Greene
5A Grundy
5A Harrison
4A Henry
4A Hickory
5A Holt
4A Howard
4A Howell
4A Iron
4A Jackson
4A Jasper
4A Jefferson
4A Johnson
5A Knox
4A Laclede
4A Lafayette
4A Lawrence
5A Lewis
4A Lincoln
5A Linn
5A Livingston
5A Macon
4A Madison
4A Maries
5A Marion
4A McDonald
5A Mercer
4A Miller
4A Mississippi
4A Moniteau
4A Monroe
4A Montgomery
4A Morgan
4A New Madrid
4A Newton
5A Nodaway
4A Oregon
4A Osage
4A Ozark
4A Pemiscot
4A Perry
4A Pettis
4A Phelps
5A Pike
4A Platte
4A Polk
4A Pulaski
5A Putnam
5A Ralls
4A Randolph
4A Ray
4A Reynolds
4A Ripley
4A Saline
5A Schuyler
5A Scotland
4A Scott
4A Shannon
5A Shelby
4A St. Charles
4A St. Clair
4A St. Francois
4A St. Louis
4A St. Louis (city)
4A Ste. Genevieve
4A Stoddard
4A Stone
5A Sullivan
4A Taney
4A Texas
4A Vernon
4A Warren
4A Washington
4A Wayne
4A Webster
5A Worth
4A Wright

MONTANA

6B (all)

NEBRASKA

5A (all)

NEVADA

5B Carson City (city)
5B Churchill
3B Clark
5B Douglas
5B Elko
5B Esmeralda
5B Eureka
5B Humboldt
5B Lander
5B Lincoln
5B Lyon
5B Mineral
5B Nye
5B Pershing
5B Storey
5B Washoe
5B White Pine

NEW HAMPSHIRE

6A Belknap
6A Carroll
5A Cheshire
6A Coos
6A Grafton
5A Hillsborough
6A Merrimack
5A Rockingham
5A Strafford
6A Sullivan

NEW JERSEY

4A Atlantic
5A Bergen
4A Burlington
4A Camden
4A Cape May
4A Cumberland
4A Essex
4A Gloucester
4A Hudson
5A Hunterdon
5A Mercer
4A Middlesex
4A Monmouth
5A Morris
4A Ocean
5A Passaic
4A Salem
5A Somerset
5A Sussex
4A Union
5A Warren

NEW MEXICO

4B Bernalillo
5B Catron
3B Chaves
4B Cibola
5B Colfax
4B Curry
4B DeBaca
3B Doña Ana
3B Eddy
4B Grant
4B Guadalupe
5B Harding
3B Hidalgo
3B Lea
4B Lincoln
5B Los Alamos
3B Luna
5B McKinley
5B Mora
3B Otero
4B Quay
5B Rio Arriba
4B Roosevelt
5B Sandoval
5B San Juan
5B San Miguel
5B Santa Fe
4B Sierra
4B Socorro

(continued)

TABLE C301.1—continued
CLIMATE ZONES, MOISTURE REGIMES, AND WARM-HUMID DESIGNATIONS BY STATE, COUNTY AND TERRITORY

5B Taos
5B Torrance
4B Union
4B Valencia

NEW YORK

5A Albany
6A Allegany
4A Bronx
6A Broome
6A Cattaraugus
5A Cayuga
5A Chautauqua
5A Chemung
6A Chenango
6A Clinton
5A Columbia
5A Cortland
6A Delaware
5A Dutchess
5A Erie
6A Essex
6A Franklin
6A Fulton
5A Genesee
5A Greene
6A Hamilton
6A Herkimer
6A Jefferson
4A Kings
6A Lewis
5A Livingston
6A Madison
5A Monroe
6A Montgomery
4A Nassau
4A New York
5A Niagara
6A Oneida
5A Onondaga
5A Ontario
5A Orange
5A Orleans
5A Oswego
6A Otsego
5A Putnam
4A Queens
5A Rensselaer
4A Richmond
5A Rockland
5A Saratoga
5A Schenectady
6A Schoharie
6A Schuyler
5A Seneca
6A Steuben
6A St. Lawrence
4A Suffolk
6A Sullivan
5A Tioga
6A Tompkins
6A Ulster
6A Warren
5A Washington
5A Wayne
4A Westchester
6A Wyoming
5A Yates

NORTH CAROLINA

4A Alamance
4A Alexander
5A Alleghany
3A Anson
5A Ashe
5A Avery
3A Beaufort
4A Bertie
3A Bladen
3A Brunswick*
4A Buncombe
4A Burke
3A Cabarrus
4A Caldwell
3A Camden
3A Carteret*
4A Caswell
4A Catawba
4A Chatham
4A Cherokee
3A Chowan
4A Clay
4A Cleveland
3A Columbus*
3A Craven
3A Cumberland
3A Currituck
3A Dare
3A Davidson
4A Davie
3A Duplin
4A Durham
3A Edgecombe
4A Forsyth
4A Franklin
3A Gaston
4A Gates
4A Graham
4A Granville
3A Greene
4A Guilford
4A Halifax
4A Harnett
4A Haywood
4A Henderson
4A Hertford
3A Hoke
3A Hyde
4A Iredell
4A Jackson
3A Johnston
3A Jones
4A Lee
3A Lenoir
4A Lincoln
4A Macon
4A Madison
3A Martin
4A McDowell
3A Mecklenburg
5A Mitchell
3A Montgomery
3A Moore
4A Nash
3A New Hanover*
4A Northampton
3A Onslow*
4A Orange
3A Pamlico
3A Pasquotank
3A Pender*
3A Perquimans
4A Person
3A Pitt
4A Polk
3A Randolph
3A Richmond
3A Robeson
4A Rockingham
3A Rowan
4A Rutherford
3A Sampson
3A Scotland
3A Stanly
4A Stokes
4A Surry
4A Swain
4A Transylvania
3A Tyrrell
3A Union
4A Vance
4A Wake
4A Warren
3A Washington
5A Watauga
3A Wayne
4A Wilkes
3A Wilson
4A Yadkin
5A Yancey

NORTH DAKOTA

6A Adams
7 Barnes
7 Benson
6A Billings
7 Bottineau
6A Bowman
7 Burke
6A Burleigh
7 Cass
7 Cavalier
6A Dickey
7 Divide
6A Dunn
7 Eddy
6A Emmons
7 Foster
6A Golden Valley
7 Grand Forks
6A Grant
7 Griggs
6A Hettinger
7 Kidder
6A LaMoure
6A Logan
7 McHenry
6A McIntosh
6A McKenzie
7 McLean
6A Mercer
6A Morton
7 Mountrail
7 Nelson
6A Oliver
7 Pembina
7 Pierce
7 Ramsey
6A Ransom
7 Renville
6A Richland
7 Rolette
6A Sargent
7 Sheridan
6A Sioux
6A Slope
6A Stark
7 Steele
7 Stutsman
7 Towner
7 Traill
7 Walsh
7 Ward
7 Wells
7 Williams

OHIO

4A Adams
5A Allen

(continued)

TABLE C301.1—continued
CLIMATE ZONES, MOISTURE REGIMES, AND WARM-HUMID DESIGNATIONS BY STATE, COUNTY AND TERRITORY

5A Ashland
5A Ashtabula
5A Athens
5A Auglaize
5A Belmont
4A Brown
5A Butler
5A Carroll
5A Champaign
5A Clark
4A Clermont
5A Clinton
5A Columbiana
5A Coshocton
5A Crawford
5A Cuyahoga
5A Darke
5A Defiance
5A Delaware
5A Erie
5A Fairfield
5A Fayette
5A Franklin
5A Fulton
4A Gallia
5A Geauga
5A Greene
5A Guernsey
4A Hamilton
5A Hancock
5A Hardin
5A Harrison
5A Henry
5A Highland
5A Hocking
5A Holmes
5A Huron
5A Jackson
5A Jefferson
5A Knox
5A Lake
4A Lawrence
5A Licking
5A Logan
5A Lorain
5A Lucas
5A Madison
5A Mahoning
5A Marion
5A Medina
5A Meigs
5A Mercer
5A Miami
5A Monroe
5A Montgomery
5A Morgan
5A Morrow
5A Muskingum
5A Noble
5A Ottawa
5A Paulding
5A Perry
5A Pickaway
4A Pike
5A Portage
5A Preble
5A Putnam
5A Richland
5A Ross
5A Sandusky
4A Scioto
5A Seneca
5A Shelby
5A Stark
5A Summit
5A Trumbull
5A Tuscarawas
5A Union
5A Van Wert
5A Vinton
5A Warren
4A Washington
5A Wayne
5A Williams
5A Wood
5A Wyandot

OKLAHOMA

3A Adair
3A Alfalfa
3A Atoka
4B Beaver
3A Beckham
3A Blaine
3A Bryan
3A Caddo
3A Canadian
3A Carter
3A Cherokee
3A Choctaw
4B Cimarron
3A Cleveland
3A Coal
3A Comanche
3A Cotton
3A Craig
3A Creek
3A Custer
3A Delaware
3A Dewey
3A Ellis
3A Garfield
3A Garvin
3A Grady
3A Grant
3A Greer
3A Harmon
3A Harper
3A Haskell
3A Hughes
3A Jackson
3A Jefferson
3A Johnston
3A Kay
3A Kingfisher
3A Kiowa
3A Latimer
3A Le Flore
3A Lincoln
3A Logan
3A Love
3A Major
3A Marshall
3A Mayes
3A McClain
3A McCurtain
3A McIntosh
3A Murray
3A Muskogee
3A Noble
3A Nowata
3A Okfuskee
3A Oklahoma
3A Okmulgee
3A Osage
3A Ottawa
3A Pawnee
3A Payne
3A Pittsburg
3A Pontotoc
3A Pottawatomie
3A Pushmataha
3A Roger Mills
3A Rogers
3A Seminole
3A Sequoyah
3A Stephens
4B Texas
3A Tillman
3A Tulsa
3A Wagoner
3A Washington
3A Washita
3A Woods
3A Woodward

OREGON

5B Baker
4C Benton
4C Clackamas
4C Clatsop
4C Columbia
4C Coos
5B Crook
4C Curry
5B Deschutes
4C Douglas
5B Gilliam
5B Grant
5B Harney
5B Hood River
4C Jackson
5B Jefferson
4C Josephine
5B Klamath
5B Lake
4C Lane
4C Lincoln
4C Linn
5B Malheur
4C Marion
5B Morrow
4C Multnomah
4C Polk
5B Sherman
4C Tillamook
5B Umatilla
5B Union
5B Wallowa
5B Wasco
4C Washington
5B Wheeler
4C Yamhill

PENNSYLVANIA

5A Adams
5A Allegheny
5A Armstrong
5A Beaver
5A Bedford
5A Berks
5A Blair
5A Bradford
4A Bucks
5A Butler
5A Cambria
6A Cameron
5A Carbon
5A Centre
4A Chester
5A Clarion
6A Clearfield
5A Clinton
5A Columbia
5A Crawford
5A Cumberland
5A Dauphin
4A Delaware
6A Elk
5A Erie
5A Fayette
5A Forest
5A Franklin
5A Fulton
5A Greene

(continued)

TABLE C301.1—continued
CLIMATE ZONES, MOISTURE REGIMES, AND WARM-HUMID DESIGNATIONS BY STATE, COUNTY AND TERRITORY

5A Huntingdon
5A Indiana
5A Jefferson
5A Juniata
5A Lackawanna
5A Lancaster
5A Lawrence
5A Lebanon
5A Lehigh
5A Luzerne
5A Lycoming
6A McKean
5A Mercer
5A Mifflin
5A Monroe
4A Montgomery
5A Montour
5A Northampton
5A Northumberland
5A Perry
4A Philadelphia
5A Pike
6A Potter
5A Schuylkill
5A Snyder
5A Somerset
5A Sullivan
6A Susquehanna
6A Tioga
5A Union
5A Venango
5A Warren
5A Washington
6A Wayne
5A Westmoreland
5A Wyoming
4A York

RHODE ISLAND

5A (all)

SOUTH CAROLINA

3A Abbeville
3A Aiken
3A Allendale*
3A Anderson
3A Bamberg*
3A Barnwell*
3A Beaufort*
3A Berkeley*
3A Calhoun
3A Charleston*
3A Cherokee
3A Chester
3A Chesterfield
3A Clarendon
3A Colleton*
3A Darlington
3A Dillon
3A Dorchester*
3A Edgefield
3A Fairfield
3A Florence
3A Georgetown*
3A Greenville
3A Greenwood
3A Hampton*
3A Horry*
3A Jasper*
3A Kershaw
3A Lancaster
3A Laurens
3A Lee
3A Lexington
3A Marion
3A Marlboro
3A McCormick
3A Newberry
3A Oconee
3A Orangeburg
3A Pickens
3A Richland
3A Saluda
3A Spartanburg
3A Sumter
3A Union
3A Williamsburg
3A York

SOUTH DAKOTA

6A Aurora
6A Beadle
5A Bennett
5A Bon Homme
6A Brookings
6A Brown
6A Brule
6A Buffalo
6A Butte
6A Campbell
5A Charles Mix
6A Clark
5A Clay
6A Codington
6A Corson
6A Custer
6A Davison
6A Day
6A Deuel
6A Dewey
5A Douglas
6A Edmunds
6A Fall River
6A Faulk
6A Grant
5A Gregory
6A Haakon
6A Hamlin
6A Hand
6A Hanson
6A Harding
6A Hughes
5A Hutchinson
6A Hyde
5A Jackson
6A Jerauld
6A Jones
6A Kingsbury
6A Lake
6A Lawrence
6A Lincoln
6A Lyman
6A Marshall
6A McCook
6A McPherson
6A Meade
5A Mellette
6A Miner
6A Minnehaha
6A Moody
6A Pennington
6A Perkins
6A Potter
6A Roberts
6A Sanborn
6A Shannon
6A Spink
6A Stanley
6A Sully
5A Todd
5A Tripp
6A Turner
5A Union
6A Walworth
5A Yankton
6A Ziebach

TENNESSEE

4A Anderson
4A Bedford
4A Benton
4A Bledsoe
4A Blount
4A Bradley
4A Campbell
4A Cannon
4A Carroll
4A Carter
4A Cheatham
3A Chester
4A Claiborne
4A Clay
4A Cocke
4A Coffee
3A Crockett
4A Cumberland
4A Davidson
4A Decatur
4A DeKalb
4A Dickson
3A Dyer
3A Fayette
4A Fentress
4A Franklin
4A Gibson
4A Giles
4A Grainger
4A Greene
4A Grundy
4A Hamblen
4A Hamilton
4A Hancock
3A Hardeman
3A Hardin
4A Hawkins
3A Haywood
3A Henderson
4A Henry
4A Hickman
4A Houston
4A Humphreys
4A Jackson
4A Jefferson
4A Johnson
4A Knox
3A Lake
3A Lauderdale
4A Lawrence
4A Lewis
4A Lincoln
4A Loudon
4A Macon
3A Madison
4A Marion
4A Marshall
4A Maury
4A McMinn
3A McNairy
4A Meigs
4A Monroe
4A Montgomery
4A Moore
4A Morgan
4A Obion
4A Overton
4A Perry
4A Pickett
4A Polk
4A Putnam
4A Rhea

(continued)

TABLE C301.1—continued
CLIMATE ZONES, MOISTURE REGIMES, AND WARM-HUMID DESIGNATIONS BY STATE, COUNTY AND TERRITORY

4A Roane
4A Robertson
4A Rutherford
4A Scott
4A Sequatchie
4A Sevier
3A Shelby
4A Smith
4A Stewart
4A Sullivan
4A Sumner
3A Tipton
4A Trousdale
4A Unicoi
4A Union
4A Van Buren
4A Warren
4A Washington
4A Wayne
4A Weakley
4A White
4A Williamson
4A Wilson

TEXAS

2A Anderson*
3B Andrews
2A Angelina*
2A Aransas*
3A Archer
4B Armstrong
2A Atascosa*
2A Austin*
4B Bailey
2B Bandera
2A Bastrop*
3B Baylor
2A Bee*
2A Bell*
2A Bexar*
3A Blanco*
3B Borden
2A Bosque*
3A Bowie*
2A Brazoria*
2A Brazos*
3B Brewster
4B Briscoe
2A Brooks*
3A Brown*
2A Burleson*
3A Burnet*
2A Caldwell*
2A Calhoun*
3B Callahan
2A Cameron*
3A Camp*
4B Carson
3A Cass*
4B Castro
2A Chambers*
2A Cherokee*
3B Childress
3A Clay
4B Cochran
3B Coke
3B Coleman
3A Collin*
3B Collingsworth
2A Colorado*
2A Comal*
3A Comanche*
3B Concho
3A Cooke
2A Coryell*
3B Cottle
3B Crane
3B Crockett
3B Crosby
3B Culberson
4B Dallam
3A Dallas*
3B Dawson
4B Deaf Smith
3A Delta
3A Denton*
2A DeWitt*
3B Dickens
2B Dimmit
4B Donley
2A Duval*
3A Eastland
3B Ector
2B Edwards
3A Ellis*
3B El Paso
3A Erath*
2A Falls*
3A Fannin
2A Fayette*
3B Fisher
4B Floyd
3B Foard
2A Fort Bend*
3A Franklin*
2A Freestone*
2B Frio
3B Gaines
2A Galveston*
3B Garza
3A Gillespie*
3B Glasscock
2A Goliad*
2A Gonzales*
4B Gray
3A Grayson
3A Gregg*
2A Grimes*
2A Guadalupe*
4B Hale
3B Hall
3A Hamilton*
4B Hansford
3B Hardeman
2A Hardin*
2A Harris*
3A Harrison*
4B Hartley
3B Haskell
2A Hays*
3B Hemphill
3A Henderson*
2A Hidalgo*
2A Hill*
4B Hockley
3A Hood*
3A Hopkins*
2A Houston*
3B Howard
3B Hudspeth
3A Hunt*
4B Hutchinson
3B Irion
3A Jack
2A Jackson*
2A Jasper*
3B Jeff Davis
2A Jefferson*
2A Jim Hogg*
2A Jim Wells*
3A Johnson*
3B Jones
2A Karnes*
3A Kaufman*
3A Kendall*
2A Kenedy*
3B Kent
3B Kerr
3B Kimble
3B King
2B Kinney
2A Kleberg*
3B Knox
3A Lamar*
4B Lamb
3A Lampasas*
2B La Salle
2A Lavaca*
2A Lee*
2A Leon*
2A Liberty*
2A Limestone*
4B Lipscomb
2A Live Oak*
3A Llano*
3B Loving
3B Lubbock
3B Lynn
2A Madison*
3A Marion*
3B Martin
3B Mason
2A Matagorda*
2B Maverick
3B McCulloch
2A McLennan*
2A McMullen*
2B Medina
3B Menard
3B Midland
2A Milam*
3A Mills*
3B Mitchell
3A Montague
2A Montgomery*
4B Moore
3A Morris*
3B Motley
3A Nacogdoches*
3A Navarro*
2A Newton*
3B Nolan
2A Nueces*
4B Ochiltree
4B Oldham
2A Orange*
3A Palo Pinto*
3A Panola*
3A Parker*
4B Parmer
3B Pecos
2A Polk*
4B Potter
3B Presidio
3A Rains*
4B Randall
3B Reagan
2B Real
3A Red River*
3B Reeves
2A Refugio*
4B Roberts
2A Robertson*
3A Rockwall*
3B Runnels
3A Rusk*
3A Sabine*
3A San Augustine*
2A San Jacinto*
2A San Patricio*

(continued)

TABLE C301.1—continued
CLIMATE ZONES, MOISTURE REGIMES, AND WARM-HUMID DESIGNATIONS BY STATE, COUNTY AND TERRITORY

3A San Saba*
3B Schleicher
3B Scurry
3B Shackelford
3A Shelby*
4B Sherman
3A Smith*
3A Somervell*
2A Starr*
3A Stephens
3B Sterling
3B Stonewall
3B Sutton
4B Swisher
3A Tarrant*
3B Taylor
3B Terrell
3B Terry
3B Throckmorton
3A Titus*
3B Tom Green
2A Travis*
2A Trinity*
2A Tyler*
3A Upshur*
3B Upton
2B Uvalde
2B Val Verde
3A Van Zandt*
2A Victoria*
2A Walker*
2A Waller*
3B Ward
2A Washington*
2B Webb
2A Wharton*
3B Wheeler
3A Wichita
3B Wilbarger
2A Willacy*
2A Williamson*
2A Wilson*
3B Winkler
3A Wise
3A Wood*
4B Yoakum
3A Young
2B Zapata
2B Zavala

UTAH

5B Beaver
6B Box Elder
6B Cache
6B Carbon
6B Daggett
5B Davis
6B Duchesne
5B Emery
5B Garfield
5B Grand
5B Iron
5B Juab
5B Kane
5B Millard
6B Morgan
5B Piute
6B Rich
5B Salt Lake
5B San Juan
5B Sanpete
5B Sevier
6B Summit
5B Tooele
6B Uintah
5B Utah
6B Wasatch
3B Washington
5B Wayne
5B Weber

VERMONT

6A (all)

VIRGINIA

4A (all)

WASHINGTON

5B Adams
5B Asotin
5B Benton
5B Chelan
4C Clallam
4C Clark
5B Columbia
4C Cowlitz
5B Douglas
6B Ferry
5B Franklin
5B Garfield
5B Grant
4C Grays Harbor
4C Island
4C Jefferson
4C King
4C Kitsap
5B Kittitas
5B Klickitat
4C Lewis
5B Lincoln
4C Mason
6B Okanogan
4C Pacific
6B Pend Oreille
4C Pierce
4C San Juan
4C Skagit
5B Skamania
4C Snohomish
5B Spokane
6B Stevens
4C Thurston
4C Wahkiakum
5B Walla Walla
4C Whatcom
5B Whitman
5B Yakima

WEST VIRGINIA

5A Barbour
4A Berkeley
4A Boone
4A Braxton
5A Brooke
4A Cabell
4A Calhoun
4A Clay
5A Doddridge
5A Fayette
4A Gilmer
5A Grant
5A Greenbrier
5A Hampshire
5A Hancock
5A Hardy
5A Harrison
4A Jackson
4A Jefferson
4A Kanawha
5A Lewis
4A Lincoln
4A Logan
5A Marion
5A Marshall
4A Mason
4A McDowell
4A Mercer
5A Mineral
4A Mingo
5A Monongalia
4A Monroe
4A Morgan
5A Nicholas
5A Ohio
5A Pendleton
4A Pleasants
5A Pocahontas
5A Preston
4A Putnam
5A Raleigh
5A Randolph
4A Ritchie
4A Roane
5A Summers
5A Taylor
5A Tucker
4A Tyler
5A Upshur
4A Wayne
5A Webster
5A Wetzel
4A Wirt
4A Wood
4A Wyoming

WISCONSIN

6A Adams
7 Ashland
6A Barron
7 Bayfield
6A Brown
6A Buffalo
7 Burnett
6A Calumet
6A Chippewa
6A Clark
6A Columbia
6A Crawford
6A Dane
6A Dodge
6A Door
7 Douglas
6A Dunn
6A Eau Claire
7 Florence
6A Fond du Lac
7 Forest
6A Grant
6A Green
6A Green Lake
6A Iowa
7 Iron
6A Jackson
6A Jefferson
6A Juneau
6A Kenosha
6A Kewaunee
6A La Crosse
6A Lafayette
7 Langlade
7 Lincoln
6A Manitowoc
6A Marathon
6A Marinette
6A Marquette
6A Menominee
6A Milwaukee
6A Monroe
6A Oconto
7 Oneida
6A Outagamie

(continued)

**TABLE C301.1—continued
CLIMATE ZONES, MOISTURE REGIMES, AND WARM-HUMID DESIGNATIONS BY STATE, COUNTY AND TERRITORY**

6A Ozaukee
6A Pepin
6A Pierce
6A Polk
6A Portage
7 Price
6A Racine
6A Richland
6A Rock
6A Rusk
6A Sauk
7 Sawyer
6A Shawano
6A Sheboygan
6A St. Croix
7 Taylor
6A Trempealeau
6A Vernon
7 Vilas
6A Walworth
7 Washburn
6A Washington
6A Waukesha
6A Waupaca
6A Waushara
6A Winnebago
6A Wood

WYOMING

6B Albany
6B Big Horn
6B Campbell
6B Carbon
6B Converse
6B Crook
6B Fremont
5B Goshen
6B Hot Springs
6B Johnson
6B Laramie
7 Lincoln
6B Natrona
6B Niobrara
6B Park
5B Platte
6B Sheridan
7 Sublette
6B Sweetwater
7 Teton
6B Uinta
6B Washakie
6B Weston

US TERRITORIES

AMERICAN SAMOA

1A (all)*

GUAM

1A (all)*

NORTHERN MARIANA ISLANDS

1A (all)*

PUERTO RICO

1A (all)*

VIRGIN ISLANDS

1A (all)*

**TABLE C301.3(1)
INTERNATIONAL CLIMATE ZONE DEFINITIONS**

MAJOR CLIMATE TYPE DEFINITIONS
Marine (C) Definition—Locations meeting all four criteria: 1. Mean temperature of coldest month between -3°C (27°F) and 18°C (65°F). 2. Warmest month mean < 22°C (72°F). 3. At least four months with mean temperatures over 10°C (50°F). 4. Dry season in summer. The month with the heaviest precipitation in the cold season has at least three times as much precipitation as the month with the least precipitation in the rest of the year. The cold season is October through March in the Northern Hemisphere and April through September in the Southern Hemisphere.
Dry (B) Definition—Locations meeting the following criteria: Not marine and $P_{in} < 0.44 \times (TF - 19.5)$ [$P_{cm} < 2.0 \times (TC + 7)$ in SI units] where: P_{in} = Annual precipitation in inches (cm) T = Annual mean temperature in °F (°C)
Moist (A) Definition—Locations that are not marine and not dry.
Warm-humid Definition—Moist (A) locations where either of the following wet-bulb temperature conditions shall occur during the warmest six consecutive months of the year: 1. 67°F (19.4°C) or higher for 3,000 or more hours; or 2. 73°F (22.8°C) or higher for 1,500 or more hours.

For SI: °C = [(°F) - 32]/1.8, 1 inch = 2.54 cm.

TABLE C301.3(2)
INTERNATIONAL CLIMATE ZONE DEFINITIONS

ZONE NUMBER	THERMAL CRITERIA	
	IP Units	SI Units
1	9000 < CDD50°F	5000 < CDD10°C
2	6300 < CDD50°F ≤ 9000	3500 < CDD10°C ≤ 5000
3A and 3B	4500 < CDD50°F ≤ 6300 AND HDD65°F ≤ 5400	2500 < CDD10°C ≤ 3500 AND HDD18°C ≤ 3000
4A and 4B	CDD50°F ≤ 4500 AND HDD65°F ≤ 5400	CDD10°C ≤ 2500 AND HDD18°C ≤ 3000
3C	HDD65°F ≤ 3600	HDD18°C ≤ 2000
4C	3600 < HDD65°F ≤ 5400	2000 < HDD18°C ≤ 3000
5	5400 < HDD65°F ≤ 7200	3000 < HDD18°C ≤ 4000
6	7200 < HDD65°F ≤ 9000	4000 < HDD18°C ≤ 5000
7	9000 < HDD65°F ≤ 12600	5000 < HDD18°C ≤ 7000
8	12600 < HDD65°F	7000 < HDD18°C

For SI: °C = [(°F) - 32]/1.8.

SECTION C302 DESIGN CONDITIONS

C302.1 Interior design conditions. The interior design temperatures used for heating and cooling load calculations shall be a maximum of 72°F (22°C) for heating and minimum of 75°F (24°C) for cooling.

SECTION C303 MATERIALS, SYSTEMS AND EQUIPMENT

C303.1 Identification. Materials, systems and equipment shall be identified in a manner that will allow a determination of compliance with the applicable provisions of this code.

C303.1.1 Building thermal envelope insulation. An *R*-value identification mark shall be applied by the manufacturer to each piece of *building thermal envelope* insulation 12 inches (305 mm) or greater in width. Alternatively, the insulation installers shall provide a certification listing the type, manufacturer and *R*-value of insulation installed in each element of the *building thermal envelope*. For blown-in or sprayed fiberglass and cellulose insulation, the initial installed thickness, settled thickness, settled *R*-value, installed density, coverage area and number of bags installed shall be *listed* on the certification. For sprayed polyurethane foam (SPF) insulation, the installed thickness of the areas covered and *R*-value of installed thickness shall be *listed* on the certification. For insulated siding, the *R*-value shall be labeled on the product's package and shall be *listed* on the certification. The insulation installer shall sign, date and post the certification in a conspicuous location on the job site.

Exception: For roof insulation installed above the deck, the *R*-value shall be labeled as required by the material standards specified in Table 1508.2 of the *International Building Code*.

C303.1.1.1 Blown-in or sprayed roof/ceiling insulation. The thickness of blown-in or sprayed fiberglass and cellulose roof/ceiling insulation shall be written in inches (mm) on markers and one or more of such markers shall be installed for every 300 square feet (28 m^2) of attic area throughout the attic space. The markers shall be affixed to the trusses or joists and marked with the minimum initial installed thickness with numbers not less than 1 inch (25 mm) in height. Each marker shall face the attic *access* opening. Spray polyurethane foam thickness and installed *R*-value shall be *listed* on certification provided by the insulation installer.

C303.1.2 Insulation mark installation. Insulating materials shall be installed such that the manufacturer's *R*-value mark is readily observable upon inspection.

C303.1.3 Fenestration product rating. *U*-factors of fenestration products shall be determined as follows:

1. For windows, doors and skylights, *U*-factor ratings shall be determined in accordance with NFRC 100.
2. Where required for garage doors and rolling doors, *U*-factor ratings shall be determined in accordance with either NFRC 100 or ANSI/DASMA 105.

U-factors shall be determined by an accredited, independent laboratory, and *labeled* and certified by the manufacturer.

Products lacking such a *labeled* *U*-factor shall be assigned a default *U*-factor from Table C303.1.3(1) or C303.1.3(2). The solar heat gain coefficient (SHGC) and *visible transmittance* (VT) of glazed fenestration products (windows, glazed doors and skylights) shall be determined in accordance with NFRC 200 by an accredited, independent laboratory, and *labeled* and certified by the manufacturer. Products lacking such a *labeled* SHGC or VT shall be assigned a default SHGC or VT from Table C303.1.3(3).

TABLE C303.1.3(1)
DEFAULT GLAZED WINDOW, GLASS DOOR AND SKYLIGHT *U*-FACTORS

FRAME TYPE	WINDOW AND GLASS DOOR		SKYLIGHT	
	Single	Double	Single	Double
Metal	1.20	0.80	2.00	1.30
Metal with Thermal Break	1.10	0.65	1.90	1.10
Nonmetal or Metal Clad	0.95	0.55	1.75	1.05
Glazed Block	0.60			

TABLE C303.1.3(2)
DEFAULT OPAQUE DOOR *U*-FACTORS

DOOR TYPE	OPAQUE *U*-FACTOR
Uninsulated Metal	1.20
Insulated Metal (Rolling)	0.90
Insulated Metal (Other)	0.60
Wood	0.50
Insulated, nonmetal edge, max 45% glazing, any glazing double pane	0.35

TABLE C303.1.3(3)
DEFAULT GLAZED FENESTRATION SHGC AND VT

	SINGLE GLAZED		DOUBLE GLAZED		GLAZED BLOCK
	Clear	Tinted	Clear	Tinted	
SHGC	0.8	0.7	0.7	0.6	0.6
VT	0.6	0.3	0.6	0.3	0.6

C303.1.4 Insulation product rating. The thermal resistance (*R*-value) of insulation shall be determined in accordance with the U.S. Federal Trade Commission *R*-value rule (CFR Title 16, Part 460) in units of h • ft^2 • °F/Btu at a mean temperature of 75°F (24°C).

C303.1.4.1 Insulated siding. The thermal resistance (*R*-value) of insulated siding shall be determined in accordance with ASTM C1363. Installation for testing shall be in accordance with the manufacturer's instructions.

C303.2 Installation. Materials, systems and equipment shall be installed in accordance with the manufacturer's instructions and the *International Building Code*.

C303.2.1 Protection of exposed foundation insulation. Insulation applied to the exterior of basement walls, crawl space walls and the perimeter of slab-on-grade floors shall have a rigid, opaque and weather-resistant protective covering to prevent the degradation of the insulation's thermal performance. The protective covering shall cover the exposed exterior insulation and extend not less than 6 inches (153 mm) below grade.

C303.2.2 Multiple layers of continuous insulation board. Where two or more layers of continuous insulation board are used in a construction assembly, the continuous insulation boards shall be installed in accordance with Section C303.2. Where the continuous insulation board manufacturer's instructions do not address installation of two or more layers, the edge joints between each layer of continuous insulation boards shall be staggered.

CHAPTER 4 [CE]

COMMERCIAL ENERGY EFFICIENCY

User note:

About this chapter: *Chapter 4 presents the paths and options for compliance with the energy efficiency provisions. Chapter 4 contains energy efficiency provisions for the building envelope, fenestration, mechanical systems, appliances, freezers and coolers, kitchen exhaust, interior and exterior lighting, water heating systems, transformers and motors.*

SECTION C401 GENERAL

C401.1 Scope. The provisions in this chapter are applicable to commercial *buildings* and their *building sites.*

C401.2 Application. Commercial buildings shall comply with one of the following:

1. The requirements of ANSI/ASHRAE/IESNA 90.1.
2. The requirements of Sections C402 through C405 and C408. In addition, commercial buildings shall comply with Section C406 and tenant spaces shall comply with Section C406.1.1.
3. The requirements of Sections C402.5, C403.2, C403.3 through C403.3.2, C403.4 through C403.4.2.3, C403.5.5, C403.7, C403.8.1 through C403.8.4, C403.10.1 through C403.10.3, C403.11, C403.12, C404, C405, C407 and C408. The building energy cost shall be equal to or less than 85 percent of the standard reference design building.

C401.2.1 Application to replacement fenestration products. Where some or all of an existing *fenestration* unit is replaced with a new *fenestration* product, including sash and glazing, the replacement *fenestration* unit shall meet the applicable requirements for *U*-factor and *SHGC* in Table C402.4.

Exception: An area-weighted average of the *U*-factor of replacement fenestration products being installed in the building for each fenestration product category listed in Table C402.4 shall be permitted to satisfy the *U*-factor requirements for each fenestration product category listed in Table C402.4. Individual fenestration products from different product categories listed in Table C402.4 shall not be combined in calculating the area-weighted average *U*-factor.

SECTION C402 BUILDING ENVELOPE REQUIREMENTS

C402.1 General (Prescriptive). Building thermal envelope assemblies for buildings that are intended to comply with the code on a prescriptive basis in accordance with the compliance path described in Item 2 of Section C401.2, shall comply with the following:

1. The opaque portions of the building thermal envelope shall comply with the specific insulation requirements of Section C402.2 and the thermal requirements of either the *R*-value-based method of Section C402.1.3; the *U*-, *C*- and *F*-factor-based method of Section C402.1.4; or the component performance alternative of Section C402.1.5.
2. Roof solar reflectance and thermal emittance shall comply with Section C402.3.
3. Fenestration in building envelope assemblies shall comply with Section C402.4.
4. Air leakage of building envelope assemblies shall comply with Section C402.5.

Alternatively, where buildings have a vertical fenestration area or skylight area exceeding that allowed in Section C402.4, the building and building thermal envelope shall comply with Section C401.2, Item 1 or Section C401.2, Item 3.

Walk-in coolers, walk-in freezers, refrigerated warehouse coolers and refrigerated warehouse freezers shall comply with Section C403.10.1 or C403.10.2.

C402.1.1 Low-energy buildings. The following low-energy buildings, or portions thereof separated from the remainder of the building by *building thermal envelope* assemblies complying with this section, shall be exempt from the *building thermal envelope* provisions of Section C402.

1. Those with a peak design rate of energy usage less than 3.4 Btu/h • ft^2 (10.7 W/m^2) or 1.0 watt per square foot (10.7 W/m^2) of floor area for space conditioning purposes.
2. Those that do not contain *conditioned space.*
3. Greenhouses.

C402.1.2 Equipment buildings. Buildings that comply with the following shall be exempt from the *building thermal envelope* provisions of this code:

1. Are separate buildings with floor area not more than 500 square feet (50 m^2).
2. Are intended to house electronic equipment with installed equipment power totaling not less than 7 watts per square foot (75 W/m^2) and not intended for human occupancy.
3. Have a heating system capacity not greater than (17,000 Btu/h) (5 kW) and a heating thermostat setpoint that is restricted to not more than 50°F (10°C).
4. Have an average wall and roof *U*-factor less than 0.200 in *Climate Zones* 1 through 5 and less than 0.120 in *Climate Zones* 6 through 8.
5. Comply with the roof solar reflectance and thermal emittance provisions for *Climate Zone* 1.

C402.1.3 Insulation component *R*-value-based method. *Building thermal envelope* opaque assemblies shall comply with the requirements of Sections C402.2 and C402.4 based on the *climate zone* specified in Chapter 3. For opaque portions of the *building thermal envelope* intended to comply on an insulation component *R-value* basis, the *R*-values for insulation shall be not less than that specified in Table C402.1.3. Commercial buildings or portions of commercial buildings enclosing *Group R* occupancies shall use the *R*-values from the "*Group R*" column of Table C402.1.3. Commercial buildings or portions of commercial buildings enclosing occupancies other than *Group R* shall use the *R*-values from the "All other" column of Table C402.1.3.

C402.1.4 Assembly *U*-factor, *C*-factor or *F*-factor-based method. Building thermal envelope opaque assemblies shall meet the requirements of Sections C402.2 and C402.4 based on the climate zone specified in Chapter 3. Building thermal envelope opaque assemblies intended to comply on an assembly *U*-, *C*- or *F*-factor basis shall have a *U*-, *C*- or *F*-factor not greater than that specified in Table C402.1.4. Commercial buildings or portions of commercial buildings enclosing *Group R* occupancies shall use the *U*-, *C*- or *F*-factor from the "*Group R*" column of Table C402.1.4. Commercial buildings or portions of commercial buildings enclosing occupancies other than *Group R* shall use the *U*-, *C*- or *F*-factor from the "All other" column of Table C402.1.4

C402.1.4.1 Thermal resistance of cold-formed steel walls. *U*-factors of walls with cold-formed steel studs shall be permitted to be determined in accordance with Equation 4-1:

$$U = 1/[R_s + (ER)] \quad \textbf{(Equation 4-1)}$$

where:

R_s = The cumulative *R*-value of the wall components along the path of heat transfer, excluding the *cavity insulation* and steel studs.

ER = The effective *R*-value of the *cavity insulation* with steel studs as specified in Table C402.1.4.1.

C402.1.5 Component performance alternative. Building envelope values and fenestration areas determined in accordance with Equation 4-2 shall be an alternative to compliance with the *U*-, *F*- and *C*-factors in Tables C402.1.4 and C402.4 and the maximum allowable fenestration areas in Section C402.4.1. *Fenestration* shall meet the applicable SHGC requirements of Section C402.4.3.

$$A + B + C + D + E \leq \text{Zero} \quad \textbf{(Equation 4-2)}$$

where:

A = Sum of the (UA Dif) values for each distinct assembly type of the building thermal envelope, other than slabs on grade and below-grade walls.

UA Dif = UA Proposed - UA Table.

UA Proposed = Proposed *U*-value × Area.

UA Table = (*U*-factor from Table C402.1.3, C402.1.4 or C402.4 × Area.

B = Sum of the (FL Dif) values for each distinct slab-on-grade perimeter condition of the building thermal envelope.

FL Dif = FL Proposed - FL Table.

FL Proposed = Proposed *F*-value × Perimeter length.

FL Table = (*F*-factor specified in Table C402.1.4) × Perimeter length.

C = Sum of the (CA Dif) values for each distinct below-grade wall assembly type of the building thermal envelope.

CA Dif = CA Proposed - CA Table.

CA Proposed = Proposed *C*-value × Area.

CA Table = (Maximum allowable *C*-factor specified in Table C402.1.4) × Area.

Where the proposed vertical glazing area is less than or equal to the maximum vertical glazing area allowed by Section C402.4.1, the value of D (Excess Vertical Glazing Value) shall be zero. Otherwise:

D = (DA × UV) - (DA × U Wall), but not less than zero.

DA = (Proposed Vertical Glazing Area) - (Vertical Glazing Area allowed by Section C402.4.1).

UA Wall = Sum of the (UA Proposed) values for each opaque assembly of the exterior wall.

TABLE C402.1.4.1
EFFECTIVE *R*-VALUES FOR STEEL STUD WALL ASSEMBLIES

NOMINAL STUD DEPTH (inches)	SPACING OF FRAMING (inches)	CAVITY *R*-VALUE (insulation)	CORRECTION FACTOR (F_c)	EFFECTIVE *R*-VALUE (ER) (Cavity *R*-Value × F_c)
$3^1/_2$	16	13	0.46	5.98
		15	0.43	6.45
$3^1/_2$	24	13	0.55	7.15
		15	0.52	7.80
6	16	19	0.37	7.03
		21	0.35	7.35
6	24	19	0.45	8.55
		21	0.43	9.03
8	16	25	0.31	7.75
	24	25	0.38	9.50

TABLE C402.1.3
OPAQUE THERMAL ENVELOPE INSULATION COMPONENT MINIMUM REQUIREMENTS, R-VALUE METHOD[a, i]

CLIMATE ZONE	1		2		3		4 EXCEPT MARINE		5 AND MARINE 4		6		7		8	
	All other	Group R	All other	Group R	All other	Group R	All other	Group R	All other	Group R	All other	Group R	All other	Group R	All other	Group R
Roofs																
Insulation entirely above roof deck	R-20ci	R-25ci	R-25ci	R-25ci	R-25ci	R-25ci	R-30ci	R-30ci	R-30ci	R-30ci	R-30ci	R-30ci	R-35ci	R-35ci	R-35ci	R-35ci
Metal buildings[b]	R-19 + R-11 LS	R-19 + R-11 LS	R-19 + R11 LS	R-19 + R-11 LS	R-19 + R-11 LS	R-19 + R-11 LS	R-19 + R-11 LS	R-19 + R-11 LS	R-19 + R-11 LS	R-19 + R-11 LS	R-25 + R-11 LS	R-25 + R-11 LS	R-30 + R-11 LS	R-30 + R-11 LS	R-30 + R-11 LS	R-30 + R-11 LS
Attic and other	R-38	R-38	R-38	R-38	R-38	R-38	R-38	R-38	R-38	R-49	R-49	R-49	R-49	R-49	R-49	R-49
Walls, above grade																
Mass[g]	R-5.7ci[c]	R-5.7ci[c]	R-5.7ci[c]	R-7.6ci	R-7.6ci	R-9.5ci	R-9.5ci	R-11.4ci	R-11.4ci	R-13.3ci	R-13.3ci	R-15.2ci	R-15.2ci	R-15.2ci	R-25ci	R-25ci
Metal building	R-13+ R-6.5ci	R-13 + R-6.5ci	R13 + R-6.5ci	R-13 + R-13ci	R-13 + R-6.5ci	R-13 + R-13ci	R-13 + R-13ci	R-13 + R-13ci	R-13 + R-13ci	R-13 + R-13ci	R-13 + R-13ci	R-13 + R-13ci	R-13 + R-13ci	R-13+ R-19.5ci	R-13 + R-13ci	R-13+ R-19.5ci
Metal framed	R-13 + R-5ci	R-13 + R-5ci	R-13 + R-5ci	R-13 + R-7.5ci	R-13 + R-7.5ci	R-13 + R-7.5ci	R-13 + R-7.5ci	R-13 + R-7.5ci	R-13 + R-7.5ci	R-13 + R-7.5ci	R-13 + R-7.5ci	R-13 + R-7.5ci	R-13 + R-7.5ci	R-13 + R-15.6ci	R-13 + R-7.5ci	R-13+ R17.5ci
Wood framed and other	R-13 + R-3.8ci or R-20	R-13 + R-3.8ci or R-20	R-13 + R-3.8ci or R-20	R-13 + R-3.8ci or R-20	R-13 + R-3.8ci or R-20	R-13 + R-3.8ci or R-20	R-13 + R-3.8ci or R-20	R-13 + R-3.8ci or R-20	R-13 + R-3.8ci or R-20	R-13 + R-7.5ci or R-20 + R-3.8ci	R-13 + R-7.5ci or R-20 + R-3.8ci	R-13 + R-7.5ci or R-20 + R-3.8ci	R-13 + R-7.5ci or R-20 + R-3.8ci	R-13 + R-7.5ci or R-20 + R-3.8ci	R13 + R-15.6ci or R-20 + R-10ci	R13 + R-15.6ci or R-20 + R-10ci
Walls, below grade																
Below-grade wall[d]	NR	NR	NR	NR	NR	NR	R-7.5ci	R-7.5ci	R-7.5ci	R-7.5ci	R-7.5ci	R-7.5ci	R-10ci	R-10ci	R-10ci	R-12.5ci
Floors																
Mass[e]	NR	NR	R-6.3ci	R-8.3ci	R-10ci	R-10ci	R-10ci	R-10.4ci	R-10ci	R-12.5ci	R-12.5ci	R-12.5ci	R-15ci	R-16.7ci	R-15ci	R-16.7ci
Joist/framing	NR	NR	R-30	R-30	R-30	R-30	R-30	R-30	R-30	R-30	R-30	R-30[f]	R-30[f]	R-30[f]	R-30[f]	R-30[f]
Slab-on-grade floors																
Unheated slabs	NR	NR	NR	NR	NR	NR	R-10 for 24″ below	R-10 for 24″ below	R-10 for 24″ below	R-10 for 24″ below	R-10 for 24″ below	R-15 for 24″ below	R-15 for 24″ below	R-15 for 24″ below	R-15 for 24″ below	R-20 for 24″ below
Heated slabs[h]	R-7.5 for 12″ below + R-5 full slab	R-7.5 for 12″ below + R-5 full slab	R-7.5 for 12″ below + R-5 full slab	R-7.5 for 12″ below + R-5 full slab	R-10 for 24″ below + R-5 full slab	R-10 for 24″ below + R-5 full slab	R-15 for 24″ below + R-5 full slab	R-15 for 24″ below + R-5 full slab	R-15 for 36″ below + R-5 full slab	R-15 for 36″ below + R-5 full slab	R-15 for 36″ below + R-5 full slab	R-20 for 48″ below + R-5 full slab	R-20 for 48″ below + R-5 full slab	R-20 for 48″ below + R-5 full slab	R-20 for 48″ below + R-5 full slab	R-20 for 48″ below + R-5 full slab
Opaque doors																
Nonswinging	R-4.75	R-4.75	R-4.75	R-4.75	R-4.75	R-4.75	R-4.75	R-4.75	R-4.75	R-4.75	R-4.75	R-4.75	R-4.75	R-4.75	R-4.75	R-4.75

For SI: 1 inch = 25.4 mm, 1 pound per square foot = 4.88 kg/m^2, 1 pound per cubic foot = 16 kg/m^3.

ci = Continuous insulation, NR = No Requirement, LS = Liner System.

a. Assembly descriptions can be found in ANSI/ASHRAE/IESNA Appendix A.

b. Where using *R*-value compliance method, a thermal spacer block shall be provided, otherwise use the *U*-factor compliance method in Table C402.1.4.

c. R-5.7ci is allowed to be substituted with concrete block walls complying with ASTM C90, ungrouted or partially grouted at 32 inches or less on center vertically and 48 inches or less on center horizontally, with ungrouted cores filled with materials having a maximum thermal conductivity of 0.44 Btu-in/h-f^2 °F.

d. Where heated slabs are below grade, below-grade walls shall comply with the exterior insulation requirements for heated slabs.

e. "Mass floors" shall be in accordance with Section C402.2.3.

f. Steel floor joist systems shall be insulated to R-38.

g. "Mass walls" shall be in accordance with Section C402.2.2.

h. The first value is for perimeter insulation and the second value is for slab insulation. Perimeter insulation is not required to extend below the bottom of the slab.

i. Not applicable to garage doors. See Table C402.1.4.

TABLE C402.1.4
OPAQUE THERMAL ENVELOPE ASSEMBLY MAXIMUM REQUIREMENTS, *U*-FACTOR METHOD[a, b]

CLIMATE ZONE	1		2		3		4 EXCEPT MARINE		5 AND MARINE 4		6		7		8	
	All other	Group R	All other	Group R	All other	Group R	All other	Group R	All other	Group R	All other	Group R	All other	Group R	All other	Group R
Roofs																
Insulation entirely above roof deck	U-0.048	U-0.039	U-0.039	U-0.039	U-0.039	U-0.039	U-0.032	U-0.032	U-0.032	U-0.032	U-0.032	U-0.032	U-0.028	U-0.028	U-0.028	U-0.028
Metal buildings	U-0.044	U-0.035	U-0.035	U-0.035	U-0.035	U-0.035	U-0.035	U-0.035	U-0.035	U-0.035	U-0.031	U-0.031	U-0.029	U-0.029	U-0.029	U-0.029
Attic and other	U-0.027	U-0.027	U-0.027	U-0.027	U-0.027	U-0.027	U-0.027	U-0.027	U-0.027	U-0.021	U-0.021	U-0.021	U-0.021	U-0.021	U-0.021	U-0.021
Walls, above grade																
Mass[g]	U-0.151	U-0.151	U-0.151	U-0.123	U-0.123	U-0.104	U-0.104	U-0.090	U-0.090	U-0.080	U-0.080	U-0.071	U-0.071	U-0.071	U-0.061	U-0.061
Metal building	U-0.079	U-0.079	U-0.079	U-0.079	U-0.079	U-0.052	U-0.052	U-0.052	U-0.052	U-0.052	U-0.052	U-0.052	U-0.052	U-0.039	U-0.052	U-0.039
Metal framed	U-0.077	U-0.077	U-0.077	U-0.064	U-0.064	U-0.064	U-0.064	U-0.064	U-0.064	U-0.064	U-0.064	U-0.064	U-0.064	U-0.052	U-0.064	U-0.045
Wood framed and other[c]	U-0.064	U-0.064	U-0.064	U-0.064	U-0.064	U-0.064	U-0.064	U-0.064	U-0.064	U-0.064	U-0.051	U-0.051	U-0.051	U-0.051	U-0.036	U-0.036
Walls, below grade																
Below-grade wall[c]	C-1.140[e]	C-1.140[e]	C-1.140[e]	C-1.140[e]	C-1.140[e]	C-1.140[e]	C-0.119	C-0.119	C-0.119	C-0.119	C-0.119	C-0.119	C-0.092	C-0.092	C-0.092	C-0.092
Floors																
Mass[d]	U-0.322[e]	U-0.322[e]	U-0.107	U-0.087	U-0.076	U-0.076	U-0.076	U-0.074	U-0.074	U-0.064	U-0.064	U-0.064	U-0.055	U-0.051	U-0.055	U-0.051
Joist/framing	U-0.066[e]	U-0.066[e]	U-0.033	U-0.033	U-0.033	U-0.033	U-0.033	U-0.033	U-0.033	U-0.033	U-0.033	U-0.033	U-0.033	U-0.033	U-0.033	U-0.033
Slab-on-grade floors																
Unheated slabs	F-0.73[e]	F-0.73[e]	F-0.73[e]	F-0.73[e]	F-0.73[e]	F-0.73[e]	F-0.54	F-0.54	F-0.54	F-0.54	F-0.54	F-0.52	F-0.40	F-0.40	F-0.40	F-0.40
Heated slabs[f]	F-1.02 0.74	F-1.02 0.74	F-1.02 0.74	F-1.02 0.74	F-0.90 0.74	F-0.90 0.74	F-0.86 0.64	F-0.86 0.64	F-0.79 0.64	F-0.79 0.64	F-0.79 0.55	F-0.69 0.55	F-0.69 0.55	F-0.69 0.55	F-0.69 0.55	F-0.69 0.55
Opaque doors																
Swinging door	U-0.61	U-0.61	U-0.61	U-0.61	U-0.61	U-0.61	U-0.61	U-0.61	U-0.37	U-0.37	U-0.37	U-0.37	U-0.37	U-0.37	U-0.37	U-0.37
Garage door <14% glazing	U-0.31	U-0.31	U-0.31	U-0.31	U-0.31	U-0.31	U-0.31	U-0.31	U-0.31	U-0.31	U-0.31	U-0.31	U-0.31	U-0.31	U-0.31	U-0.31

For SI: 1 pound per square foot = 4.88 kg/m^2, 1 pound per cubic foot = 16 kg/m^3.
ci = Continuous insulation, NR = No Requirement, LS = Liner System.

a. Where assembly *U*-factors, *C*-factors, and *F*-factors are established in ANSI/ASHRAE/IESNA 90.1 Appendix A, such opaque assemblies shall be a compliance alternative where those values meet the criteria of this table, and provided that the construction, excluding the cladding system on walls, complies with the appropriate construction details from ANSI/ASHRAE/ISNEA 90.1 Appendix A.
b. Where *U*-factors have been established by testing in accordance with ASTM C1363, such opaque assemblies shall be a compliance alternative where those values meet the criteria of this table. The *R*-value of continuous insulation shall be permitted to be added to or subtracted from the original tested design.
c. Where heated slabs are below grade, below-grade walls shall comply with the *U*-factor requirements for above-grade mass walls.
d. "Mass floors" shall be in accordance with Section C402.2.3.
e. These *C*-, *F*- and *U*-factors are based on assemblies that are not required to contain insulation.
f. The first value is for perimeter insulation and the second value is for full slab insulation.
g. "Mass walls" shall be in accordance with Section C402.2.2.

U Wall = Area-weighted average *U*-value of all above-grade wall assemblies.

UAV = Sum of the (UA Proposed) values for each vertical glazing assembly.

UV = UAV/total vertical glazing area.

Where the proposed skylight area is less than or equal to the skylight area allowed by Section C402.4.1, the value of E (Excess Skylight Value) shall be zero. Otherwise:

E = (EA × US) - (EA × U Roof), but not less than zero.

EA = (Proposed Skylight Area) - (Allowable Skylight Area as specified in Section C402.4.1).

U Roof = Area-weighted average *U*-value of all roof assemblies.

UAS = Sum of the (UA Proposed) values for each skylight assembly.

US = UAS/total skylight area.

C402.2 Specific building thermal envelope insulation requirements (Prescriptive). Insulation in building thermal envelope opaque assemblies shall comply with Sections C402.2.1 through C402.2.7 and Table C402.1.3.

C402.2.1 Roof assembly. The minimum thermal resistance (*R*-value) of the insulating material installed either between the roof framing or continuously on the roof assembly shall be as specified in Table C402.1.3, based on construction materials used in the roof assembly. Insulation installed on a suspended ceiling having removable ceiling tiles shall not be considered as part of the minimum thermal resistance of the roof insulation. Continuous insulation board shall be installed in not less than 2 layers and the edge joints between each layer of insulation shall be staggered.

Exceptions:

1. Continuously insulated roof assemblies where the thickness of insulation varies 1 inch (25 mm) or less and where the area-weighted *U*-factor is equivalent to the same assembly with the *R*-value specified in Table C402.1.3.
2. Where tapered insulation is used with insulation entirely above deck, the *R*-value where the insulation thickness varies 1 inch (25 mm) or less from the minimum thickness of tapered insulation shall comply with the *R*-value specified in Table C402.1.3.
3. Two layers of insulation are not required where insulation tapers to the roof deck, such as at roof drains.

C402.2.1.1 Skylight curbs. Skylight curbs shall be insulated to the level of roofs with insulation entirely above the deck or R-5, whichever is less.

Exception: Unit skylight curbs included as a component of a skylight listed and labeled in accordance with NFRC 100 shall not be required to be insulated.

C402.2.2 Above-grade walls. The minimum thermal resistance (*R*-value) of materials installed in the wall cavity between framing members and continuously on the walls shall be as specified in Table C402.1.3, based on framing type and construction materials used in the wall assembly. The *R*-value of integral insulation installed in concrete masonry units shall not be used in determining compliance with Table C402.1.3 except as otherwise noted in the table. In determining compliance with Table C402.1.4, the use of the *U*-factor of concrete masonry units with integral insulation shall be permitted.

"Mass walls" where used as a component in the thermal envelope of a building shall comply with one of the following:

1. Weigh not less than 35 pounds per square foot (171 kg/m^2) of wall surface area.
2. Weigh not less than 25 pounds per square foot (122 kg/m^2) of wall surface area where the material weight is not more than 120 pcf (1900 kg/m^3).
3. Have a heat capacity exceeding 7 $Btu/ft^2 \cdot °F$ (144 $kJ/m^2 \cdot K$).
4. Have a heat capacity exceeding 5 $Btu/ft^2 \cdot °F$ (103 $kJ/m^2 \cdot K$), where the material weight is not more than 120 pcf (1900 kg/m^3).

C402.2.3 Floors. The thermal properties (component *R*-values or assembly *U*-, *C*- or *F*-factors) of floor assemblies over outdoor air or unconditioned space shall be as specified in Table C402.1.3 or C402.1.4 based on the construction materials used in the floor assembly. Floor framing *cavity insulation* or structural slab insulation shall be installed to maintain permanent contact with the underside of the subfloor decking or structural slabs.

"Mass floors" where used as a component of the thermal envelope of a building shall provide one of the following weights:

1. 35 pounds per square foot (171 kg/m^2) of floor surface area.
2. 25 pounds per square foot (122 kg/m^2) of floor surface area where the material weight is not more than 120 pounds per cubic foot (1923 kg/m^3).

Exceptions:

1. The floor framing *cavity insulation* or structural slab insulation shall be permitted to be in contact with the top side of sheathing or continuous insulation installed on the bottom side of floor assemblies where combined with insulation that meets or exceeds the minimum *R*-value in Table C402.1.3 for "Metal framed" or "Wood framed and other" values for "Walls, Above Grade" and extends from the bottom to the top of all perimeter floor framing or floor assembly members.
2. Insulation applied to the underside of concrete floor slabs shall be permitted an airspace of not more than 1 inch (25 mm) where it turns up and is in contact with the underside of the floor under walls associated with the *building thermal envelope*.

C402.2.4 Slabs-on-grade perimeter insulation. Where the slab on grade is in contact with the ground, the minimum thermal resistance (*R*-value) of the insulation around the perimeter of unheated or heated slab-on-grade floors designed in accordance with the *R*-value method of Section C402.1.3 shall be as specified in Table C402.1.3. The perimeter insulation shall be placed on the outside of the foundation or on the inside of the foundation wall. The perimeter insulation shall extend downward from the top of the slab for the minimum distance shown in the table or to the top of the footing, whichever is less, or downward to not less than the bottom of the slab and then horizontally to the interior or exterior for the total distance shown in the table. Insulation extending away from the building shall be protected by pavement or by not less than of 10 inches (254 mm) of soil.

> **Exception:** Where the slab-on-grade floor is greater than 24 inches (61 mm) below the finished exterior grade, perimeter insulation is not required.

C402.2.5 Below-grade walls. The *C*-factor for the below-grade exterior walls shall be in accordance with Table C402.1.4. The *R*-value of the insulating material installed continuously within or on the below-grade exterior walls of the building envelope shall be in accordance with Table C402.1.3. The *C*-factor or *R*-value required shall extend to a depth of not less than 10 feet (3048 mm) below the outside finished ground level, or to the level of the lowest floor of the conditioned space enclosed by the below-grade wall, whichever is less.

C402.2.6 Insulation of radiant heating systems. *Radiant heating system* panels, and their associated components that are installed in interior or exterior assemblies shall be insulated to an *R*-value of not less than R-3.5 on all surfaces not facing the space being heated. *Radiant heating system* panels that are installed in the *building thermal envelope* shall be separated from the exterior of the building or unconditioned or exempt spaces by not less than the *R*-value of insulation installed in the opaque assembly in which they are installed or the assembly shall comply with Section C402.1.4.

> **Exception:** Heated slabs on grade insulated in accordance with Section C402.2.4.

C402.2.7 Airspaces. Where the thermal properties of airspaces are used to comply with this code in accordance with Section C401.2, such airspaces shall be enclosed in an unventilated cavity constructed to minimize airflow into and out of the enclosed airspace. Airflow shall be deemed minimized where the enclosed airspace is located on the interior side of the continuous air barrier and is bounded on all sides by building components.

> **Exception:** The thermal resistance of airspaces located on the exterior side of the continuous air barrier and adjacent to and behind the exterior wall-covering material shall be determined in accordance with ASTM C1363 modified with an airflow entering the bottom and exiting the top of the airspace at an air movement rate of not less than 70 mm/second.

C402.3 Roof solar reflectance and thermal emittance. Low-sloped roofs directly above cooled conditioned spaces in *Climate Zones* 1, 2 and 3 shall comply with one or more of the options in Table C402.3.

> **Exceptions:** The following roofs and portions of roofs are exempt from the requirements of Table C402.3:
>
> 1. Portions of the roof that include or are covered by the following:
>   - 1.1. Photovoltaic systems or components.
>   - 1.2. Solar air or water-heating systems or components.
>   - 1.3. Roof gardens or landscaped roofs.
>   - 1.4. Above-roof decks or walkways.
>   - 1.5. Skylights.
>   - 1.6. HVAC systems and components, and other opaque objects mounted above the roof.
> 2. Portions of the roof shaded during the peak sun angle on the summer solstice by permanent features of the building or by permanent features of adjacent buildings.
> 3. Portions of roofs that are ballasted with a minimum stone ballast of 17 pounds per square foot [74 kg/m^2] or 23 psf [117 kg/m^2] pavers.
> 4. Roofs where not less than 75 percent of the roof area complies with one or more of the exceptions to this section.

C402.3.1 Aged roof solar reflectance. Where an aged solar reflectance required by Section C402.3 is not available, it shall be determined in accordance with Equation 4-3.

$$R_{aged} = [0.2 + 0.7(R_{initial} - 0.2)] \quad \textbf{(Equation 4-3)}$$

where:

R_{aged} = The aged solar reflectance.

$R_{initial}$ = The initial solar reflectance determined in accordance with CRRC-S100.

C402.4 Fenestration (Prescriptive). Fenestration shall comply with Sections C402.4.1 through C402.4.5 and Table C402.4. Daylight responsive controls shall comply with this section and Section C405.2.3.1.

C402.4.1 Maximum area. The vertical fenestration area, not including opaque doors and opaque spandrel panels, shall be not greater than 30 percent of the gross above-grade wall area. The skylight area shall be not greater than 3 percent of the gross roof area.

C402.4.1.1 Increased vertical fenestration area with daylight responsive controls. In *Climate Zones* 1 through 6, not more than 40 percent of the gross above-grade wall area shall be vertical fenestration, provided that all of the following requirements are met:

1. In buildings not greater than two stories above grade, not less than 50 percent of the net floor area is within a *daylight zone.*

TABLE C402.3
MINIMUM ROOF REFLECTANCE AND EMITTANCE OPTIONS[a]

Three-year-aged solar reflectance[b] of 0.55 and 3-year aged thermal emittance[c] of 0.75
Three-year-aged solar reflectance index[d] of 64

a. The use of area-weighted averages to comply with these requirements shall be permitted. Materials lacking 3-year-aged tested values for either solar reflectance or thermal emittance shall be assigned both a 3-year-aged solar reflectance in accordance with Section C402.3.1 and a 3-year-aged thermal emittance of 0.90.

b. Aged solar reflectance tested in accordance with ASTM C1549, ASTM E903 or ASTM E1918 or CRRC-S100.

c. Aged thermal emittance tested in accordance with ASTM C1371 or ASTM E408 or CRRC-S100.

d. Solar reflectance index (SRI) shall be determined in accordance with ASTM E1980 using a convection coefficient of 2.1 Btu/h • ft² •°F (12W/m² • K). Calculation of aged SRI shall be based on aged tested values of solar reflectance and thermal emittance.

TABLE C402.4
BUILDING ENVELOPE FENESTRATION MAXIMUM *U*-FACTOR AND SHGC REQUIREMENTS

CLIMATE ZONE	1		2		3		4 EXCEPT MARINE		5 AND MARINE 4		6		7		8	
Vertical fenestration																
***U*-factor**																
Fixed fenestration	0.50		0.50		0.46		0.38		0.38		0.36		0.29		0.29	
Operable fenestration	0.65		0.65		0.60		0.45		0.45		0.43		0.37		0.37	
Entrance doors	1.10		0.83		0.77		0.77		0.77		0.77		0.77		0.77	
SHGC																
Orientation[a]	SEW	N	SEW	N	SEW	N	SEW	N	SEW	N	SEW	N	SEW	N	SEW	N
PF < 0.2	0.25	0.33	0.25	0.33	0.25	0.33	0.36	0.48	0.38	0.51	0.40	0.53	0.45	NR	0.45	N
0.2 ≤ PF < 0.5	0.30	0.37	0.30	0.37	0.30	0.37	0.43	0.53	0.46	0.56	0.48	0.58	NR	NR	NR	NR
PF ≥ 0.5	0.40	0.40	0.40	0.40	0.40	0.40	0.58	0.58	0.61	0.61	0.64	0.64	NR	NR	NR	NR
Skylights																
U-factor	0.75		0.65		0.55		0.50		0.50		0.50		0.50		0.50	
SHGC	0.35		0.35		0.35		0.40		0.40		0.40		NR		NR	

NR = No Requirement, PF = Projection Factor.

a. "N" indicates vertical fenestration oriented within 45 degrees of true north. "SEW" indicates orientations other than "N." For buildings in the southern hemisphere, reverse south and north. Buildings located at less than 23.5 degrees latitude shall use SEW for all orientations.

2. In buildings three or more stories above grade, not less than 25 percent of the net floor area is within a *daylight zone.*
3. *Daylight responsive controls* complying with Section C405.2.3.1 are installed in *daylight zones.*
4. Visible transmittance (VT) of vertical fenestration is not less than 1.1 times solar heat gain coefficient (SHGC).

Exception: Fenestration that is outside the scope of NFRC 200 is not required to comply with Item 4.

C402.4.1.2 Increased skylight area with daylight responsive controls. The skylight area shall be not more than 6 percent of the roof area provided that *daylight responsive controls* complying with Section C405.2.3.1 are installed in *toplit zones*.

C402.4.2 Minimum skylight fenestration area. In an enclosed space greater than 2,500 square feet (232 m²) in floor area, directly under a roof with not less than 75 percent of the ceiling area with a ceiling height greater than 15 feet (4572 mm), and used as an office, lobby, atrium, concourse, corridor, storage space, gymnasium/exercise center, convention center, automotive service area, space where manufacturing occurs, nonrefrigerated warehouse, retail store, distribution/sorting area, transportation depot or workshop, the total *toplit daylight zone* shall be not less than half the floor area and shall provide one of the following:

1. A minimum skylight area to *toplit daylight zone* of not less than 3 percent where all skylights have a VT of not less than 0.40 as determined in accordance with Section C303.1.3.
2. A minimum skylight effective aperture of not less than 1 percent, determined in accordance with Equation 4-4.

$$\text{Skylight Effective Aperture} = \frac{0.85 \times \text{Skylight Area} \times \text{Skylight VT} \times \text{WF}}{\text{Toplit Zone}}$$

(Equation 4-4)

where:

Skylight area = Total fenestration area of skylights.

Skylight VT = Area weighted average visible transmittance of skylights.

WF = Area weighted average well factor, where well factor is 0.9 if light well depth is less than 2 feet (610 mm), or 0.7 if light well depth is 2 feet (610 mm) or greater.

Light well depth = Measure vertically from the underside of the lowest point of the skylight glazing to the ceiling plane under the skylight.

Exception: Skylights above *daylight zones* of enclosed spaces are not required in:

1. Buildings in *Climate Zones* 6 through 8.
2. Spaces where the designed *general lighting* power densities are less than 0.5 W/ft^2 (5.4 W/m^2).
3. Areas where it is documented that existing structures or natural objects block direct beam sunlight on not less than half of the roof over the enclosed area for more than 1,500 daytime hours per year between 8 a.m. and 4 p.m.
4. Spaces where the *daylight zone* under rooftop monitors is greater than 50 percent of the enclosed space floor area.
5. Spaces where the total area minus the area of sidelight *daylight zones* is less than 2,500 square feet (232 m^2), and where the lighting is controlled in accordance with Section C405.2.3.

C402.4.2.1 Lighting controls in toplit daylight zones. *Daylight responsive controls* complying with Section C405.2.3.1 shall be provided to control all electric lights within *toplit zones*.

C402.4.2.2 Haze factor. Skylights in office, storage, automotive service, manufacturing, nonrefrigerated warehouse, retail store and distribution/sorting area spaces shall have a glazing material or diffuser with a haze factor greater than 90 percent when tested in accordance with ASTM D1003.

Exception: Skylights designed and installed to exclude direct sunlight entering the occupied space by the use of fixed or automated baffles or the geometry of skylight and light well.

C402.4.3 Maximum *U*-factor and SHGC. The maximum *U*-factor and solar heat gain coefficient (SHGC) for fenestration shall be as specified in Table C402.4.

The window projection factor shall be determined in accordance with Equation 4-5.

$$PF = A/B \quad \textbf{(Equation 4-5)}$$

where:

PF = Projection factor (decimal).

A = Distance measured horizontally from the farthest continuous extremity of any overhang, eave or permanently attached shading device to the vertical surface of the glazing.

B = Distance measured vertically from the bottom of the glazing to the underside of the overhang, eave or permanently attached shading device.

Where different windows or glass doors have different PF values, they shall each be evaluated separately.

C402.4.3.1 Increased skylight SHGC. In *Climate Zones* 1 through 6, skylights shall be permitted a maximum SHGC of 0.60 where located above *daylight zones* provided with *daylight responsive controls*.

C402.4.3.2 Increased skylight *U*-factor. Where skylights are installed above *daylight zones* provided with *daylight responsive controls*, a maximum *U*-factor of 0.9 shall be permitted in *Climate Zones* 1 through 3 and a maximum *U*-factor of 0.75 shall be permitted in *Climate Zones* 4 through 8.

C402.4.3.3 Dynamic glazing. Where dynamic glazing is intended to satisfy the SHGC and VT requirements of Table C402.4, the ratio of the higher to lower labeled SHGC shall be greater than or equal to 2.4, and the *dynamic glazing* shall be automatically controlled to modulate the amount of solar gain into the space in multiple steps. Dynamic glazing shall be considered separately from other fenestration, and area-weighted averaging with other fenestration that is not dynamic glazing shall not be permitted.

Exception: Dynamic glazing is not required to comply with this section where both the lower and higher labeled SHGC already comply with the requirements of Table C402.4.

C402.4.3.4 Area-weighted *U*-factor. An area-weighted average shall be permitted to satisfy the *U*-factor requirements for each fenestration product category listed in Table C402.4. Individual fenestration products from different fenestration product categories listed in Table C402.4 shall not be combined in calculating area-weighted average *U*-factor.

C402.4.4 Daylight zones. Daylight zones referenced in Sections C402.4.1.1 through C402.4.3.2 shall comply with Sections C405.2.3.2 and C405.2.3.3, as applicable. Daylight zones shall include *toplit zones* and sidelit zones.

C402.4.5 Doors. Opaque swinging doors shall comply with Table C402.1.4. Opaque nonswinging doors shall comply with Table C402.1.3. Opaque doors shall be considered as part of the gross area of above-grade walls that are part of the building *thermal envelope*. Other doors shall comply with the provisions of Section C402.4.3 for vertical fenestration.

C402.5 Air leakage—thermal envelope (Mandatory). The *thermal envelope* of buildings shall comply with Sections C402.5.1 through C402.5.8, or the building *thermal envelope* shall be tested in accordance with ASTM E 779 at a pressure differential of 0.3 inch water gauge (75 Pa) or an equivalent method approved by the code official and deemed to comply with the provisions of this section when the tested air leakage rate of the building thermal envelope is not greater than 0.40 cfm/ft^2 (2.0 L/s • m^2). Where compliance is based on such testing, the building shall also comply with Sections C402.5.5, C402.5.6 and C402.5.7.

C402.5.1 Air barriers. A continuous air barrier shall be provided throughout the building thermal envelope. The air barriers shall be permitted to be located on the inside or outside of the building envelope, located within the assemblies composing the envelope, or any combination thereof. The air barrier shall comply with Sections C402.5.1.1 and C402.5.1.2.

Exception: Air barriers are not required in buildings located in *Climate Zone* 2B.

C402.5.1.1 Air barrier construction. The *continuous air barrier* shall be constructed to comply with the following:

1. The air barrier shall be continuous for all assemblies that are the thermal envelope of the building and across the joints and assemblies.
2. Air barrier joints and seams shall be sealed, including sealing transitions in places and changes in materials. The joints and seals shall be securely installed in or on the joint for its entire length so as not to dislodge, loosen or otherwise impair its ability to resist positive and negative pressure from wind, stack effect and mechanical ventilation.
3. Penetrations of the air barrier shall be caulked, gasketed or otherwise sealed in a manner compatible with the construction materials and location. Sealing shall allow for expansion, contraction and mechanical vibration. Joints and seams associated with penetrations shall be sealed in the same manner or taped. Sealing materials shall be securely installed around the penetration so as not to dislodge, loosen or otherwise impair the penetrations' ability to resist positive and negative pressure from wind, stack effect and mechanical ventilation. Sealing of concealed fire sprinklers, where required, shall be in a manner that is recommended by the manufacturer. Caulking or other adhesive sealants shall not be used to fill voids between fire sprinkler cover plates and walls or ceilings.
4. Recessed lighting fixtures shall comply with Section C402.5.8. Where similar objects are installed that penetrate the air barrier, provisions shall be made to maintain the integrity of the air barrier.

C402.5.1.2 Air barrier compliance options. A continuous air barrier for the opaque building envelope shall comply with Section C402.5.1.2.1 or C402.5.1.2.2.

C402.5.1.2.1 Materials. Materials with an air permeability not greater than 0.004 cfm/ft^2 (0.02 L/s • m^2) under a pressure differential of 0.3 inch water gauge (75 Pa) when tested in accordance with ASTM E2178 shall comply with this section. Materials in Items 1 through 16 shall be deemed to comply with this section, provided that joints are sealed and materials are installed as air barriers in accordance with the manufacturer's instructions.

1. Plywood with a thickness of not less than $^3/_8$ inch (10 mm).
2. Oriented strand board having a thickness of not less than $^3/_8$ inch (10 mm).
3. Extruded polystyrene insulation board having a thickness of not less than $^1/_2$ inch (12.7 mm).
4. Foil-back polyisocyanurate insulation board having a thickness of not less than $^1/_2$ inch (12.7 mm).
5. Closed-cell spray foam having a minimum density of 1.5 pcf (2.4 kg/m^3) and having a thickness of not less than 1$^1/_2$ inches (38 mm).
6. Open-cell spray foam with a density between 0.4 and 1.5 pcf (0.6 and 2.4 kg/m^3) and having a thickness of not less than 4.5 inches (113 mm).
7. Exterior or interior gypsum board having a thickness of not less than $^1/_2$ inch (12.7 mm).
8. Cement board having a thickness of not less than $^1/_2$ inch (12.7 mm).
9. Built-up roofing membrane.
10. Modified bituminous roof membrane.
11. Fully adhered single-ply roof membrane.
12. A Portland cement/sand parge, or gypsum plaster having a thickness of not less than $^5/_8$ inch (15.9 mm).
13. Cast-in-place and precast concrete.
14. Fully grouted concrete block masonry.
15. Sheet steel or aluminum.
16. Solid or hollow masonry constructed of clay or shale masonry units.

C402.5.1.2.2 Assemblies. Assemblies of materials and components with an average air leakage not greater than 0.04 cfm/ft^2 (0.2 L/s • m^2) under a pressure differential of 0.3 inch of water gauge (w.g.)(75 Pa) when tested in accordance with ASTM E2357, ASTM E1677 or ASTM E283 shall comply with this section. Assemblies listed in Items 1 through 3 shall be deemed to comply, provided that joints are sealed and the requirements of Section C402.5.1.1 are met.

1. Concrete masonry walls coated with either one application of block filler or two applications of a paint or sealer coating.
2. Masonry walls constructed of clay or shale masonry units with a nominal width of 4 inches (102 mm) or more.
3. A Portland cement/sand parge, stucco or plaster not less than $^1/_2$ inch (12.7 mm) in thickness.

C402.5.2 Air leakage of fenestration. The air leakage of fenestration assemblies shall meet the provisions of Table C402.5.2. Testing shall be in accordance with the applicable reference test standard in Table C402.5.2 by an accredited, independent testing laboratory and *labeled* by the manufacturer.

Exceptions:

1. Field-fabricated fenestration assemblies that are sealed in accordance with Section C402.5.1.
2. Fenestration in buildings that comply with the testing alternative of Section C402.5 are not required to meet the air leakage requirements in Table C402.5.2.

TABLE C402.5.2
MAXIMUM AIR LEAKAGE RATE
FOR FENESTRATION ASSEMBLIES

FENESTRATION ASSEMBLY	MAXIMUM RATE (CFM/FT²)	TEST PROCEDURE
Windows	0.20[a]	AAMA/WDMA/ CSA101/I.S.2/A440 or NFRC 400
Sliding doors	0.20[a]	
Swinging doors	0.20[a]	
Skylights – with condensation weepage openings	0.30	
Skylights – all other	0.20[a]	
Curtain walls	0.06	NFRC 400 or ASTM E283 at 1.57 psf (75 Pa)
Storefront glazing	0.06	
Commercial glazed swinging entrance doors	1.00	
Power-operated sliding doors and power-operated folding doors	1.00	
Revolving doors	1.00	
Garage doors	0.40	ANSI/DASMA 105, NFRC 400, or ASTM E283 at 1.57 psf (75 Pa)
Rolling doors	1.00	
High-speed doors	1.30	

For SI: 1 cubic foot per minute = 0.47 L/s, 1 square foot = 0.093 m².

a. The maximum rate for windows, sliding and swinging doors, and skylights is permitted to be 0.3 cfm per square foot of fenestration or door area when tested in accordance with AAMA/WDMA/CSA101/I.S.2/A440 at 6.24 psf (300 Pa).

C402.5.3 Rooms containing fuel-burning appliances. In *Climate Zones* 3 through 8, where combustion air is supplied through openings in an exterior wall to a room or space containing a space-conditioning fuel-burning appliance, one of the following shall apply:

1. The room or space containing the appliance shall be located outside of the *building thermal envelope*.
2. The room or space containing the appliance shall be enclosed and isolated from conditioned spaces inside the building thermal envelope. Such rooms shall comply with all of the following:
   2.1. The walls, floors and ceilings that separate the enclosed room or space from conditioned spaces shall be insulated to be not less than equivalent to the insulation requirement of below-grade walls as specified in Table C402.1.3 or C402.1.4.
   2.2. The walls, floors and ceilings that separate the enclosed room or space from conditioned spaces shall be sealed in accordance with Section C402.5.1.1.
   2.3. The doors into the enclosed room or space shall be shall be fully gasketed.
   2.4. Water lines and ducts in the enclosed room or space shall be insulated in accordance with Section C403.
   2.5. Where an air duct supplying combustion air to the enclosed room or space passes through conditioned space, the duct shall be insulated to an *R*-value of not less than R-8.

Exception: Fireplaces and stoves complying with Sections 901 through 905 of the *International Mechanical Code*, and Section 2111.14 of the *International Building Code*.

C402.5.4 Doors and *access* openings to shafts, chutes, stairways and elevator lobbies. Doors and *access* openings from conditioned space to shafts, chutes stairways and elevator lobbies not within the scope of the fenestration assemblies covered by Section C402.5.2 shall be gasketed, weatherstripped or sealed.

Exceptions:

1. Door openings required to comply with Section 716 of the *International Building Code.*
2. Doors and door openings required to comply with UL 1784 by the *International Building Code.*

C402.5.5 Air intakes, exhaust openings, stairways and shafts. Stairway enclosures, elevator shaft vents and other outdoor air intakes and exhaust openings integral to the building envelope shall be provided with dampers in accordance with Section C403.7.7.

C402.5.6 Loading dock weatherseals. Cargo door openings and loading door openings shall be equipped with weatherseals that restrict infiltration and provide direct contact along the top and sides of vehicles that are parked in the doorway.

C402.5.7 Vestibules. Building entrances shall be protected with an enclosed vestibule, with all doors opening into and out of the vestibule equipped with self-closing devices. Vestibules shall be designed so that in passing through the vestibule it is not necessary for the interior and exterior doors to open at the same time. The installation of one or more revolving doors in the building entrance shall not eliminate the requirement that a vestibule be provided on any doors adjacent to revolving doors.

Exceptions: Vestibules are not required for the following:

1. Buildings in *Climate Zones* 1 and 2.
2. Doors not intended to be used by the public, such as doors to mechanical or electrical equipment rooms, or intended solely for employee use.
3. Doors opening directly from a *sleeping unit* or dwelling unit.
4. Doors that open directly from a space less than 3,000 square feet (298 m²) in area.
5. Revolving doors.
6. Doors used primarily to facilitate vehicular movement or material handling and adjacent personnel doors.
7. Doors that have an air curtain with a velocity of not less than 6.56 feet per second (2 m/s) at the floor that have been tested in accordance with ANSI/AMCA 220 and installed in accordance with the manufacturer's instructions. Manual or

automatic controls shall be provided that will operate the air curtain with the opening and closing of the door. Air curtains and their controls shall comply with Section C408.2.3.

C402.5.8 Recessed lighting. Recessed luminaires installed in the *building thermal envelope* shall be all of the following:

1. IC-rated.
2. Labeled as having an air leakage rate of not more 2.0 cfm (0.944 L/s) when tested in accordance with ASTM E283 at a 1.57 psf (75 Pa) pressure differential.
3. Sealed with a gasket or caulk between the housing and interior wall or ceiling covering.

SECTION C403 BUILDING MECHANICAL SYSTEMS

C403.1 General. Mechanical systems and equipment serving the building heating, cooling, ventilating or refrigerating needs shall comply with this section.

C403.1.1 Calculation of heating and cooling loads. Design loads associated with heating, ventilating and air conditioning of the building shall be determined in accordance with ANSI/ASHRAE/ACCA Standard 183 or by an *approved* equivalent computational procedure using the design parameters specified in Chapter 3. Heating and cooling loads shall be adjusted to account for load reductions that are achieved where energy recovery systems are utilized in the HVAC system in accordance with the ASHRAE *HVAC Systems and Equipment Handbook* by an approved equivalent computational procedure.

C403.2 System design (Mandatory). Mechanical systems shall be designed to comply with Sections C403.2.1 and C403.2.2. Where elements of a building's mechanical systems are addressed in Sections C403.3 through C403.12, such elements shall comply with the applicable provisions of those sections.

C403.2.1 Zone isolation required (Mandatory). HVAC systems serving *zones* that are over 25,000 square feet (2323 m^2) in floor area or that span more than one floor and are designed to operate or be occupied nonsimultaneously shall be divided into isolation areas. Each isolation area shall be equipped with *isolation devices* and controls configured to automatically shut off the supply of conditioned air and outdoor air to and exhaust air from the isolation area. Each isolation area shall be controlled independently by a device meeting the requirements of Section C403.4.2.2. Central systems and plants shall be provided with controls and devices that will allow system and equipment operation for any length of time while serving only the smallest isolation area served by the system or plant.

Exceptions:

1. Exhaust air and outdoor air connections to isolation areas where the fan system to which they connect is not greater than 5,000 cfm (2360 L/s).
2. Exhaust airflow from a single isolation area of less than 10 percent of the design airflow of the exhaust system to which it connects.
3. Isolation areas intended to operate continuously or intended to be inoperative only when all other isolation areas in a *zone* are inoperative.

C403.2.2 Ventilation (Mandatory). Ventilation, either natural or mechanical, shall be provided in accordance with Chapter 4 of the *International Mechanical Code*. Where mechanical ventilation is provided, the system shall provide the capability to reduce the outdoor air supply to the minimum required by Chapter 4 of the *International Mechanical Code*.

C403.3 Heating and cooling equipment efficiencies (Mandatory). Heating and cooling equipment installed in mechanical systems shall be sized in accordance with Section C403.3.1 and shall be not less efficient in the use of energy than as specified in Section C403.3.2.

C403.3.1 Equipment sizing (Mandatory). The output capacity of heating and cooling equipment shall be not greater than that of the smallest available equipment size that exceeds the loads calculated in accordance with Section C403.1.1. A single piece of equipment providing both heating and cooling shall satisfy this provision for one function with the capacity for the other function as small as possible, within available equipment options.

Exceptions:

1. Required standby equipment and systems provided with controls and devices that allow such systems or equipment to operate automatically only when the primary equipment is not operating.
2. Multiple units of the same equipment type with combined capacities exceeding the design load and provided with controls that are configured to sequence the operation of each unit based on load.

C403.3.2 HVAC equipment performance requirements (Mandatory). Equipment shall meet the minimum efficiency requirements of Tables C403.3.2(1) through C403.3.2(9) when tested and rated in accordance with the applicable test procedure. Plate-type liquid-to-liquid heat exchangers shall meet the minimum requirements of Table C403.3.2(10). The efficiency shall be verified through certification under an approved certification program or, where a certification program does not exist, the equipment efficiency ratings shall be supported by data furnished by the manufacturer. Where multiple rating conditions or performance requirements are provided, the equipment shall satisfy all stated requirements. Where components, such as indoor or outdoor coils, from different manufacturers are used, calculations and supporting data shall be furnished by the designer that demonstrates that the combined efficiency of the specified components meets the requirements herein.

C403.3.2.1 Water-cooled centrifugal chilling packages (Mandatory). Equipment not designed for operation at AHRI Standard 550/590 test conditions of 44°F

(7°C) leaving chilled-water temperature and 2.4 gpm/ton evaporator fluid flow and 85°F (29°C) entering condenser water temperature with 3 gpm/ton (0.054 I/s • kW) condenser water flow shall have maximum full-load kW/ton (FL) and part-load ratings requirements adjusted using Equations 4-6 and 4-7.

$$FL_{adj} = FL/K_{adj} \quad \textbf{(Equation 4-6)}$$

$$PLV_{adj} = IPLV/K_{adj} \quad \textbf{(Equation 4-7)}$$

where:

K_{adj} = A × B

FL = Full-load kW/ton value as specified in Table C403.3.2(7).

FL_{adj} = Maximum full-load kW/ton rating, adjusted for nonstandard conditions.

$IPLV$ = Value as specified in Table C403.3.2(7).

PLV_{adj} = Maximum *NPLV* rating, adjusted for nonstandard conditions.

A = $0.00000014592 \times (LIFT)^4 - 0.0000346496 \times (LIFT)^3 + 0.00314196 \times (LIFT)^2 - 0.147199 \times (LIFT) + 3.9302$

B = $0.0015 \times L_{vg}E_{vap} + 0.934$

$LIFT$ = $L_{vg}Cond - L_{vg}E_{vap}$

$L_{vg}Cond$ = Full-load condenser leaving fluid temperature (°F).

$L_{vg}E_{vap}$ = Full-load evaporator leaving temperature (°F).

The FL_{adj} and PLV_{adj} values are only applicable for centrifugal chillers meeting all of the following full-load design ranges:

1. Minimum evaporator leaving temperature: 36°F.
2. Maximum condenser leaving temperature: 115°F.
3. 20°F ≤ LIFT ≤ 80°F.

C403.3.2.2 Positive displacement (air- and water-cooled) chilling packages (Mandatory). Equipment with a leaving fluid temperature higher than 32°F (0°C) and water-cooled positive displacement chilling packages with a condenser leaving fluid temperature below 115°F (46°C) shall meet the requirements of Table C403.3.2(7) when tested or certified with water at standard rating conditions, in accordance with the referenced test procedure.

C403.3.3 Hot gas bypass limitation. Cooling systems shall not use hot gas bypass or other evaporator pressure control systems unless the system is designed with multiple steps of unloading or continuous capacity modulation. The capacity of the hot gas bypass shall be limited as indicated in Table C403.3.3, as limited by Section C403.5.1.

TABLE C403.3.3
MAXIMUM HOT GAS BYPASS CAPACITY

RATED CAPACITY	MAXIMUM HOT GAS BYPASS CAPACITY (% of total capacity)
≤ 240,000 Btu/h	50
> 240,000 Btu/h	25

For SI: 1 British thermal unit per hour = 0.2931 W.

C403.3.4 Boiler turndown. *Boiler systems* with design input of greater than 1,000,000 Btu/h (293 kW) shall comply with the turndown ratio specified in Table C403.3.4.

The system turndown requirement shall be met through the use of multiple single-input boilers, one or more *modulating boilers* or a combination of single-input and *modulating boilers*.

TABLE C403.3.4
BOILER TURNDOWN

BOILER SYSTEM DESIGN INPUT (Btu/h)	MINIMUM TURNDOWN RATIO
≥ 1,000,000 and less than or equal to 5,000,000	3 to 1
> 5,000,000 and less than or equal to 10,000,000	4 to 1
> 10,000,000	5 to 1

For SI: 1 British thermal unit per hour = 0.2931 W.

C403.4 Heating and cooling system controls (Mandatory). Each heating and cooling system shall be provided with controls in accordance with Sections C403.4.1 through C403.4.5.

C403.4.1 Thermostatic controls (Mandatory). The supply of heating and cooling energy to each *zone* shall be controlled by individual thermostatic controls capable of responding to temperature within the *zone*. Where humidification or dehumidification or both is provided, not fewer than one humidity control device shall be provided for each humidity control system.

Exception: Independent perimeter systems that are designed to offset only building envelope heat losses, gains or both serving one or more perimeter *zones* also served by an interior system provided that both of the following conditions are met:

1. The perimeter system includes not fewer than one thermostatic control *zone* for each building exposure having exterior walls facing only one orientation (within ± 45 degrees) (0.8 rad) for more than 50 contiguous feet (15 240 mm).
2. The perimeter system heating and cooling supply is controlled by thermostats located within the *zones* served by the system.

C403.4.1.1 Heat pump supplementary heat (Mandatory). Heat pumps having supplementary electric resistance heat shall have controls that, except during defrost, prevent supplementary heat operation where the heat pump can provide the heating load.

TABLE C403.3.2(1)
MINIMUM EFFICIENCY REQUIREMENTS:
ELECTRICALLY OPERATED UNITARY AIR CONDITIONERS AND CONDENSING UNITS

EQUIPMENT TYPE	SIZE CATEGORY	HEATING SECTION TYPE	SUBCATEGORY OR RATING CONDITION	MINIMUM EFFICIENCY	TEST PROCEDURE[a]
Air conditioners, air cooled	< 65,000 Btu/h[b]	All	Split System	13.0 SEER	AHRI 210/240
			Single Package	14.0 SEER	
Through-the-wall (air cooled)	≤ 30,000 Btu/h[b]	All	Split system	12.0 SEER	
			Single Package	12.0 SEER	
Small-duct high-velocity (air cooled)	< 65,000 Btu/h[b]	All	Split System	11.0 SEER	
Air conditioners, air cooled	≥ 65,000 Btu/h and < 135,000 Btu/h	Electric Resistance (or None)	Split System and Single Package	11.2 EER 12.8 IEER	AHRI 340/360
		All other	Split System and Single Package	11.0 EER 12.6 IEER	
	≥ 135,000 Btu/h and < 240,000 Btu/h	Electric Resistance (or None)	Split System and Single Package	11.0 EER 12.4 IEER	
		All other	Split System and Single Package	10.8 EER 12.2 IEER	
	≥ 240,000 Btu/h and < 760,000 Btu/h	Electric Resistance (or None)	Split System and Single Package	10.0 EER 11.6 IEER	
		All other	Split System and Single Package	9.8 EER 11.4 IEER	
	≥ 760,000 Btu/h	Electric Resistance (or None)	Split System and Single Package	9.7 EER 11.2 IEER	
		All other	Split System and Single Package	9.5 EER 11.0 IEER	
Air conditioners, water cooled	< 65,000 Btu/h[b]	All	Split System and Single Package	12.1 EER 12.3 IEER	AHRI 210/240
	≥ 65,000 Btu/h and < 135,000 Btu/h	Electric Resistance (or None)	Split System and Single Package	12.1 EER 13.9 IEER	AHRI 340/360
		All other	Split System and Single Package	11.9 EER 13.7 IEER	
	≥ 135,000 Btu/h and < 240,000 Btu/h	Electric Resistance (or None)	Split System and Single Package	12.5 EER 13.9 IEER	
		All other	Split System and Single Package	12.3 EER 13.7 IEER	
	≥ 240,000 Btu/h and < 760,000 Btu/h	Electric Resistance (or None)	Split System and Single Package	12.4 EER 13.6 IEER	
		All other	Split System and Single Package	12.2 EER 13.4 IEER	
	≥ 760,000 Btu/h	Electric Resistance (or None)	Split System and Single Package	12.2 EER 13.5 IEER	
		All other	Split System and Single Package	12.0 EER 13.3 IEER	

(continued)

TABLE C403.3.2(1)—continued
MINIMUM EFFICIENCY REQUIREMENTS:
ELECTRICALLY OPERATED UNITARY AIR CONDITIONERS AND CONDENSING UNITS

EQUIPMENT TYPE	SIZE CATEGORY	HEATING SECTION TYPE	SUB-CATEGORY OR RATING CONDITION	MINIMUM EFFICIENCY	TEST PROCEDURE[a]
Air conditioners, evaporatively cooled	< 65,000 Btu/h[b]	All	Split System and Single Package	12.1 EER 12.3 IEER	AHRI 210/240
	≥ 65,000 Btu/h and < 135,000 Btu/h	Electric Resistance (or None)	Split System and Single Package	12.1 EER 12.3 IEER	AHRI 340/360
		All other	Split System and Single Package	11.9 EER 12.1 IEER	
	≥ 135,000 Btu/h and < 240,000 Btu/h	Electric Resistance (or None)	Split System and Single Package	12.0 EER 12.2 IEER	
		All other	Split System and Single Package	11.8 EER 12.0 IEER	
	≥ 240,000 Btu/h and < 760,000 Btu/h	Electric Resistance (or None)	Split System and Single Package	11.9 EER 12.1 IEER	
		All other	Split System and Single Package	11.7 EER 11.9 IEER	
	≥ 760,000 Btu/h	Electric Resistance (or None)	Split System and Single Package	11.7 EER 11.9 IEER	
		All other	Split System and Single Package	11.5 EER 11.7 IEER	
Condensing units, air cooled	≥ 135,000 Btu/h	—	—	10.5 EER 11.8 IEER	AHRI 365
Condensing units, water cooled	≥ 135,000 Btu/h	—	—	13.5 EER 14.0 IEER	
Condensing units, evaporatively cooled	≥ 135,000 Btu/h	—	—	13.5 EER 14.0 IEER	

For SI: 1 British thermal unit per hour = 0.2931 W.

a. Chapter 6 contains a complete specification of the referenced test procedure, including the reference year version of the test procedure.

b. Single-phase, air-cooled air conditioners less than 65,000 Btu/h are regulated by NAECA. SEER values are those set by NAECA.

TABLE C403.3.2(2)
MINIMUM EFFICIENCY REQUIREMENTS:
ELECTRICALLY OPERATED UNITARY AND APPLIED HEAT PUMPS

EQUIPMENT TYPE	SIZE CATEGORY	HEATING SECTION TYPE	SUBCATEGORY OR RATING CONDITION	MINIMUM EFFICIENCY	TEST PROCEDURE[a]
Air cooled (cooling mode)	< 65,000 Btu/h[b]	All	Split System	14.0 SEER	AHRI 210/240
			Single Package	14.0 SEER	
Through-the-wall, air cooled	≤ 30,000 Btu/h[b]	All	Split System	12.0 SEER	
			Single Package	12.0 SEER	
Single-duct high-velocity air cooled	< 65,000 Btu/h[b]	All	Split System	11.0 SEER	
Air cooled (cooling mode)	≥ 65,000 Btu/h and < 135,000 Btu/h	Electric Resistance (or None)	Split System and Single Package	11.0 EER 12.0 IEER	AHRI 340/360
		All other	Split System and Single Package	10.8 EER 11.8 IEER	
	≥ 135,000 Btu/h and < 240,000 Btu/h	Electric Resistance (or None)	Split System and Single Package	10.6 EER 11.6 IEER	
		All other	Split System and Single Package	10.4 EER 11.4 IEER	
	≥ 240,000 Btu/h	Electric Resistance (or None)	Split System and Single Package	9.5 EER 10.6 IEER	
		All other	Split System and Single Package	9.3 EER 9.4 IEER	
Water to Air: Water Loop (cooling mode)	< 17,000 Btu/h	All	86°F entering water	12.2 EER	ISO 13256-1
	≥ 17,000 Btu/h and < 65,000 Btu/h	All	86°F entering water	13.0 EER	
	≥ 65,000 Btu/h and < 135,000 Btu/h	All	86°F entering water	13.0 EER	
Water to Air: Ground Water (cooling mode)	< 135,000 Btu/h	All	59°F entering water	18.0 EER	ISO 13256-1
Brine to Air: Ground Loop (cooling mode)	< 135,000 Btu/h	All	77°F entering water	14.1 EER	ISO 13256-1
Water to Water: Water Loop (cooling mode)	< 135,000 Btu/h	All	86°F entering water	10.6 EER	ISO 13256-2
Water to Water: Ground Water (cooling mode)	< 135,000 Btu/h	All	59°F entering water	16.3 EER	
Brine to Water: Ground Loop (cooling mode)	< 135,000 Btu/h	All	77°F entering fluid	12.1 EER	

(continued)

TABLE C403.3.2(2)—continued
MINIMUM EFFICIENCY REQUIREMENTS:
ELECTRICALLY OPERATED UNITARY AND APPLIED HEAT PUMPS

EQUIPMENT TYPE	SIZE CATEGORY	HEATING SECTION TYPE	SUBCATEGORY OR RATING CONDITION	MINIMUM EFFICIENCY	TEST PROCEDURE[a]
Air cooled (heating mode)	< 65,000 Btu/h[b]	—	Split System	8.2 HSPF	AHRI 210/240
		—	Single Package	8.0 HSPF	
Through-the-wall, (air cooled, heating mode)	≤ 30,000 Btu/h[b] (cooling capacity)	—	Split System	7.4 HSPF	
		—	Single Package	7.4 HSPF	
Small-duct high velocity (air cooled, heating mode)	< 65,000 Btu/h[b]	—	Split System	6.8 HSPF	
Air cooled (heating mode)	≥ 65,000 Btu/h and < 135,000 Btu/h (cooling capacity)	—	47°F db/43°F wb outdoor air	3.3 COP	AHRI 340/360
			17°F db/15°F wb outdoor air	2.25 COP	
	≥ 135,000 Btu/h (cooling capacity)	—	47°F db/43°F wb outdoor air	3.2 COP	
			17°F db/15°F wb outdoor air	2.05 COP	
Water to Air: Water Loop (heating mode)	< 135,000 Btu/h (cooling capacity)	—	68°F entering water	4.3 COP	ISO 13256-1
Water to Air: Ground Water (heating mode)	< 135,000 Btu/h (cooling capacity)	—	50°F entering water	3.7 COP	
Brine to Air: Ground Loop (heating mode)	< 135,000 Btu/h (cooling capacity)	—	32°F entering fluid	3.2 COP	
Water to Water: Water Loop (heating mode)	< 135,000 Btu/h (cooling capacity)	—	68°F entering water	3.7 COP	ISO 13256-2
Water to Water: Ground Water (heating mode)	< 135,000 Btu/h (cooling capacity)	—	50°F entering water	3.1 COP	
Brine to Water: Ground Loop (heating mode)	< 135,000 Btu/h (cooling capacity)	—	32°F entering fluid	2.5 COP	

For SI: 1 British thermal unit per hour = 0.2931 W, °C = [(°F) - 32]/1.8.

a. Chapter 6 contains a complete specification of the referenced test procedure, including the reference year version of the test procedure.

b. Single-phase, air-cooled heat pumps less than 65,000 Btu/h are regulated by NAECA. SEER and HSPF values are those set by NAECA.

TABLE C403.3.2(3)
MINIMUM EFFICIENCY REQUIREMENTS:
ELECTRICALLY OPERATED PACKAGED TERMINAL AIR CONDITIONERS, PACKAGED TERMINAL HEAT PUMPS, SINGLE-PACKAGE VERTICAL AIR CONDITIONERS, SINGLE VERTICAL HEAT PUMPS, ROOM AIR CONDITIONERS AND ROOM AIR-CONDITIONER HEAT PUMPS

EQUIPMENT TYPE	SIZE CATEGORY (INPUT)	SUBCATEGORY OR RATING CONDITION	MINIMUM EFFICIENCY	TEST PROCEDURE[a]
PTAC (cooling mode) new construction	All Capacities	95°F db outdoor air	14.0 – (0.300 × Cap/1000) EER	AHRI 310/380
PTAC (cooling mode) replacements[b]	All Capacities	95°F db outdoor air	10.9 - (0.213 × Cap/1000) EER	
PTHP (cooling mode) new construction	All Capacities	95°F db outdoor air	14.0 - (0.300 × Cap/1000) EER	
PTHP (cooling mode) replacements[b]	All Capacities	95°F db outdoor air	10.8 - (0.213 × Cap/1000) EER	
PTHP (heating mode) new construction	All Capacities	—	3.2 - (0.026 × Cap/1000) COP	
PTHP (heating mode) replacements[b]	All Capacities	—	2.9 - (0.026 × Cap/1000) COP	
SPVAC (cooling mode)	< 65,000 Btu/h	95°F db/ 75°F wb outdoor air	9.0 EER	AHRI 390
	≥ 65,000 Btu/h and < 135,000 Btu/h	95°F db/ 75°F wb outdoor air	8.9 EER	
	≥ 135,000 Btu/h and < 240,000 Btu/h	95°F db/ 75°F wb outdoor air	8.6 EER	
SPVHP (cooling mode)	< 65,000 Btu/h	95°F db/ 75°F wb outdoor air	9.0 EER	
	≥ 65,000 Btu/h and < 135,000 Btu/h	95°F db/ 75°F wb outdoor air	8.9 EER	
	≥ 135,000 Btu/h and < 240,000 Btu/h	95°F db/ 75°F wb outdoor air	8.6 EER	
SPVHP (heating mode)	< 65,000 Btu/h	47°F db/ 43°F wb outdoor air	3.0 COP	AHRI 390
	≥ 65,000 Btu/h and < 135,000 Btu/h	47°F db/ 43°F wb outdoor air	3.0 COP	
	≥ 135,000 Btu/h and < 240,000 Btu/h	47°F db/ 75°F wb outdoor air	2.9 COP	
Room air conditioners, with louvered sides	< 6,000 Btu/h	—	11.0 CEER	ANSI/ AHAM RAC-1
	≥ 6,000 Btu/h and < 8,000 Btu/h	—	11.0 CEER	
	≥ 8,000 Btu/h and < 14,000 Btu/h	—	10.9 CEER	
	≥ 14,000 Btu/h and < 20,000 Btu/h	—	10.7 CEER	
	≥ 20,000 Btu/h and ≤ 25,000 Btu/h	—	9.4 CEER	
	> 25,000 Btu/h	—	9.0 CEER	
Room air conditioners, without louvered sides	< 6,000 Btu/h	—	10.0 CEER	
	≥ 6,000 Btu/h and < 8,000 Btu/h	—	10.0 CEER	
	≥ 8,000 Btu/h and < 11,000 Btu/h	—	9.6 CEER	
	≥ 11,000 Btu/h and < 14,000 Btu/h	—	9.5 CEER	
	≥ 14,000 Btu/h and < 20,000 Btu/h	—	9.3 CEER	
	≥ 20,000 Btu/h	—	9.4 CEER	
Room air-conditioner heat pumps with louvered sides	< 20,000 Btu/h	—	9.8 CEER	
	≥ 20,000 Btu/h	—	9.3 CEER	
Room air-conditioner heat pumps without louvered sides	< 14,000 Btu/h	—	9.3 CEER	
	≥ 14,000 Btu/h	—	8.7 CEER	

(continued)

TABLE C403.3.2(3)—continued
MINIMUM EFFICIENCY REQUIREMENTS:
ELECTRICALLY OPERATED PACKAGED TERMINAL AIR CONDITIONERS,
PACKAGED TERMINAL HEAT PUMPS, SINGLE-PACKAGE VERTICAL AIR CONDITIONERS,
SINGLE VERTICAL HEAT PUMPS, ROOM AIR CONDITIONERS AND ROOM AIR-CONDITIONER HEAT PUMPS

EQUIPMENT TYPE	SIZE CATEGORY (INPUT)	SUBCATEGORY OR RATING CONDITION	MINIMUM EFFICIENCY	TEST PROCEDURE[a]
Room air conditioner casement only	All capacities	—	9.5 CEER	ANSI/ AHAM RAC-1
Room air conditioner casement-slider	All capacities	—	10.4 CEER	

For SI: 1 British thermal unit per hour = 0.2931 W, °C = [(°F) - 32]/1.8, wb = wet bulb, db = dry bulb.

"Cap" = The rated cooling capacity of the project in Btu/h. Where the unit's capacity is less than 7000 Btu/h, use 7000 Btu/h in the calculation. Where the unit's capacity is greater than 15,000 Btu/h, use 15,000 Btu/h in the calculations.

a. Chapter 6 contains a complete specification of the referenced test procedure, including the referenced year version of the test procedure.

b. Replacement unit shall be factory labeled as follows: "MANUFACTURED FOR REPLACEMENT APPLICATIONS ONLY: NOT TO BE INSTALLED IN NEW CONSTRUCTION PROJECTS." Replacement efficiencies apply only to units with existing sleeves less than 16 inches (406 mm) in height and less than 42 inches (1067 mm) in width.

TABLE C403.3.2(4)
WARM-AIR FURNACES AND COMBINATION WARM-AIR FURNACES/AIR-CONDITIONING UNITS, WARM-AIR DUCT FURNACES AND UNIT HEATERS, MINIMUM EFFICIENCY REQUIREMENTS

EQUIPMENT TYPE	SIZE CATEGORY (INPUT)	SUBCATEGORY OR RATING CONDITION	MINIMUM EFFICIENCY[d, e]	TEST PROCEDURE[a]
Warm-air furnaces, gas fired	< 225,000 Btu/h	—	80% AFUE or 80%E_t[c]	DOE 10 CFR Part 430 or ANSI Z21.47
	≥ 225,000 Btu/h	Maximum capacity[c]	80%E_t[f]	ANSI Z21.47
Warm-air furnaces, oil fired	< 225,000 Btu/h	—	83% AFUE or 80%E_t[c]	DOE 10 CFR Part 430 or UL 727
	≥ 225,000 Btu/h	Maximum capacity[b]	81%E_t[g]	UL 727
Warm-air duct furnaces, gas fired	All capacities	Maximum capacity[b]	80%E_c	ANSI Z83.8
Warm-air unit heaters, gas fired	All capacities	Maximum capacity[b]	80%E_c	ANSI Z83.8
Warm-air unit heaters, oil fired	All capacities	Maximum capacity[b]	80%E_c	UL 731

For SI: 1 British thermal unit per hour = 0.2931 W.

a. Chapter 6 contains a complete specification of the referenced test procedure, including the referenced year version of the test procedure.

b. Minimum and maximum ratings as provided for and allowed by the unit's controls.

c. Combination units not covered by the National Appliance Energy Conservation Act of 1987 (NAECA) (3-phase power or cooling capacity greater than or equal to 65,000 Btu/h [19 kW]) shall comply with either rating.

d. E_t = Thermal efficiency. See test procedure for detailed discussion.

e. E_c = Combustion efficiency (100% less flue losses). See test procedure for detailed discussion.

f. E_c = Combustion efficiency. Units shall also include an IID, have jackets not exceeding 0.75 percent of the input rating, and have either power venting or a flue damper. A vent damper is an acceptable alternative to a flue damper for those furnaces where combustion air is drawn from the conditioned space.

g. E_t = Thermal efficiency. Units shall also include an IID, have jacket losses not exceeding 0.75 percent of the input rating, and have either power venting or a flue damper. A vent damper is an acceptable alternative to a flue damper for those furnaces where combustion air is drawn from the conditioned space.

TABLE C403.3.2(5)
MINIMUM EFFICIENCY REQUIREMENTS: GAS- AND OIL-FIRED BOILERS

EQUIPMENT TYPE[a]	SUBCATEGORY OR RATING CONDITION	SIZE CATEGORY (INPUT)	MINIMUM EFFICIENCY[d, e]	TEST PROCEDURE
Boilers, hot water	Gas-fired	< 300,000 Btu/h[f, g]	82% AFUE	10 CFR Part 430
		≥ 300,000 Btu/h and ≤ 2,500,000 Btu/h[b]	80% E_t	10 CFR Part 431
		> 2,500,000 Btu/h[a]	82% E_c	
	Oil-fired[c]	< 300,000 Btu/h[g]	84% AFUE	10 CFR Part 430
		≥ 300,000 Btu/h and ≤ 2,500,000 Btu/h[b]	82% E_t	10 CFR Part 431
		> 2,500,000 Btu/h[a]	84% E_c	
Boilers, steam	Gas-fired	< 300,000 Btu/h[f]	80% AFUE	10 CFR Part 430
	Gas-fired- all, except natural draft	≥ 300,000 Btu/h and ≤ 2,500,000 Btu/h[b]	79% E_t	10 CFR Part 431
		> 2,500,000 Btu/h[a]	79% E_t	
	Gas-fired-natural draft	≥ 300,000 Btu/h and ≤ 2,500,000 Btu/h[b]	77% E_t	
		> 2,500,000 Btu/h[a]	77% E_t	
	Oil-fired[c]	< 300,000 Btu/h	82% AFUE	10 CFR Part 430
		≥ 300,000 Btu/h and ≤ 2,500,000 Btu/h[b]	81% E_t	10 CFR Part 431
		> 2,500,000 Btu/h[a]	81% E_t	

For SI: 1 British thermal unit per hour = 0.2931 W.

a. These requirements apply to boilers with rated input of 8,000,000 Btu/h or less that are not packaged boilers and to all packaged boilers. Minimum efficiency requirements for boilers cover all capacities of packaged boilers.
b. Maximum capacity – minimum and maximum ratings as provided for and allowed by the unit's controls.
c. Includes oil-fired (residual).
d. E_c = Combustion efficiency (100 percent less flue losses).
e. E_t = Thermal efficiency. See referenced standard for detailed information.
f. Boilers shall not be equipped with a constant-burning ignition pilot.
g. A boiler not equipped with a tankless domestic water heating coil shall be equipped with an automatic means for adjusting the temperature of the water such that an incremental change in inferred heat load produces a corresponding incremental change in the temperature of the water supplied.

TABLE C403.3.2(6)
MINIMUM EFFICIENCY REQUIREMENTS:
CONDENSING UNITS, ELECTRICALLY OPERATED

EQUIPMENT TYPE	SIZE CATEGORY	MINIMUM EFFICIENCY[b]	TEST PROCEDURE[a]
Condensing units, air cooled	≥ 135,000 Btu/h	10.1 EER 11.2 IPLV	AHRI 365
Condensing units, water or evaporatively cooled	≥ 135,000 Btu/h	13.1 EER 13.1 IPLV	

For SI: 1 British thermal unit per hour = 0.2931 W.

a. Chapter 6 contains a complete specification of the referenced test procedure, including the referenced year version of the test procedure.
b. IPLVs are only applicable to equipment with capacity modulation.

TABLE C403.3.2(7)
WATER CHILLING PACKAGES – EFFICIENCY REQUIREMENTS[a, b, d]

EQUIPMENT TYPE	SIZE CATEGORY	UNITS	BEFORE 1/1/2015		AS OF 1/1/2015		TEST PROCEDURE[c]
			Path A	Path B	Path A	Path B	
Air-cooled chillers	< 150 Tons	EER (Btu/W)	≥ 9.562 FL	NA[c]	≥ 10.100 FL	≥ 9.700 FL	AHRI 550/ 590
			≥ 12.500 IPLV		≥ 13.700 IPLV	≥ 15,800 IPLV	
	≥ 150 Tons		≥ 9.562 FL	NA[c]	≥ 10.100 FL	≥ 9.700 FL	
			≥ 12.500 IPLV		≥ 14.000 IPLV	≥ 16.100 IPLV	
Air cooled without condenser, electrically operated	All capacities	EER (Btu/W)	Air-cooled chillers without condenser shall be rated with matching condensers and complying with air-cooled chiller efficiency requirements.				
Water cooled, electrically operated positive displacement	< 75 Tons	kW/ton	≤ 0.780 FL	≤ 0.800 FL	≤ 0.750 FL	≤ 0.780 FL	
			≤ 0.630 IPLV	≤ 0.600 IPLV	≤ 0.600 IPLV	≤ 0.500 IPLV	
	≥ 75 tons and < 150 tons		≤ 0.775 FL	≤ 0.790 FL	≤ 0.720 FL	≤ 0.750 FL	
			≤ 0.615 IPLV	≤ 0.586 IPLV	≤ 0.560 IPLV	≤ 0.490 IPLV	
	≥ 150 tons and < 300 tons		≤ 0.680 FL	≤ 0.718 FL	≤ 0.660 FL	≤ 0.680 FL	
			≤ 0.580 IPLV	≤ 0.540 IPLV	≤ 0.540 IPLV	≤ 0.440 IPLV	
	≥ 300 tons and < 600 tons		≤ 0.620 FL	≤ 0.639 FL	≤ 0.610 FL	≤ 0.625 FL	
			≤ 0.540 IPLV	≤ 0.490 IPLV	≤ 0.520 IPLV	≤ 0.410 IPLV	
	≥ 600 tons		≤ 0.620 FL	≤ 0.639 FL	≤ 0.560 FL	≤ 0.585 FL	
			≤ 0.540 IPLV	≤ 0.490 IPLV	≤ 0.500 IPLV	≤ 0.380 IPLV	
Water cooled, electrically operated centrifugal	< 150 Tons	kW/ton	≤ 0.634 FL	≤ 0.639 FL	≤ 0.610 FL	≤ 0.695 FL	
			≤ 0.596 IPLV	≤ 0.450 IPLV	≤ 0.550 IPLV	≤ 0.440 IPLV	
	≥ 150 tons and < 300 tons		≤ 0.634 FL	≤ 0.639 FL	≤ 0.610 FL	≤ 0.635 FL	
			≤ 0.596 IPLV	≤ 0.450 IPLV	≤ 0.550 IPLV	≤ 0.400 IPLV	
	≥ 300 tons and < 400 tons		≤ 0.576 FL	≤ 0.600 FL	≤ 0.560 FL	≤ 0.595 FL	
			≤ 0.549 IPLV	≤ 0.400 IPLV	≤ 0.520 IPLV	≤ 0.390 IPLV	
	≥ 400 tons and < 600 tons		≤ 0.576 FL	≤ 0.600 FL	≤ 0.560 FL	≤ 0.585 FL	
			≤ 0.549 IPLV	≤ 0.400 IPLV	≤ 0.500 IPLV	≤ 0.380 IPLV	
	≥ 600 Tons		≤ 0.570 FL	≤ 0.590 FL	≤ 0.560 FL	≤ 0.585 FL	
			≤ 0.539 IPLV	≤ 0.400 IPLV	≤ 0.500 IPLV	≤ 0.380 IPLV	
Air cooled, absorption, single effect	All capacities	COP	≥ 0.600 FL	NA[c]	≥ 0.600 FL	NA[c]	AHRI 560
Water cooled absorption, single effect	All capacities	COP	≥ 0.700 FL	NA[c]	≥ 0.700 FL	NA[c]	
Absorption, double effect, indirect fired	All capacities	COP	≥ 1.000 FL	NA[c]	≥ 1.000 FL	NA[c]	
			≥ 1.050 IPLV		≥ 1.050 IPLV		
Absorption double effect direct fired	All capacities	COP	≥ 1.000 FL	NA[c]	≥ 1.000 FL	NA[c]	
			≥ 1.000 IPLV		≥ 1.050 IPLV		

a. The requirements for centrifugal chiller shall be adjusted for nonstandard rating conditions in accordance with Section C403.3.2.1 and are only applicable for the range of conditions listed in Section C403.3.2.1. The requirements for air-cooled, water-cooled positive displacement and absorption chillers are at standard rating conditions defined in the reference test procedure.

b. Both the full-load and IPLV requirements shall be met or exceeded to comply with this standard. Where there is a Path B, compliance can be with either Path A or Path B for any application.

c. NA means the requirements are not applicable for Path B and only Path A can be used for compliance.

d. FL represents the full-load performance requirements and IPLV the part-load performance requirements.

TABLE C403.3.2(8)
MINIMUM EFFICIENCY REQUIREMENTS:
HEAT REJECTION EQUIPMENT

EQUIPMENT TYPE[a]	TOTAL SYSTEM HEAT REJECTION CAPACITY AT RATED CONDITIONS	SUBCATEGORY OR RATING CONDITION[i]	PERFORMANCE REQUIRED[b, c, d, g, h]	TEST PROCEDURE[e, f]
Propeller or axial fan open-circuit cooling towers	All	95°F entering water 85°F leaving water 75°F entering wb	≥ 40.2 gpm/hp	CTI ATC-105 and CTI STD-201 RS
Centrifugal fan open-circuit cooling towers	All	95°F entering water 85°F leaving water 75°F entering wb	≥ 20.0 gpm/hp	CTI ATC-105 and CTI STD-201 RS
Propeller or axial fan closed-circuit cooling towers	All	102°F entering water 90°F leaving water 75°F entering wb	≥ 16.1 gpm/hp	CTI ATC-105S and CTI STD-201 RS
Centrifugal fan closed- circuit cooling towers	All	102°F entering water 90°F leaving water 75°F entering wb	≥ 7.0 gpm/hp	CTI ATC-105S and CTI STD-201 RS
Propeller or axial fan evaporative condensers	All	Ammonia Test Fluid 140°F entering gas temperature 96.3°F condensing temperature 75°F entering wb	≥ 134,000 Btu/h × hp	CTI ATC-106
Centrifugal fan evaporative condensers	All	Ammonia Test Fluid 140°F entering gas temperature 96.3°F condensing temperature 75°F entering wb	≥110,000 Btu/h × hp	CTI ATC-106
Propeller or axial fan evaporative condensers	All	R-507A Test Fluid 165°F entering gas temperature 105°F condensing temperature 75°F entering wb	≥ 157,000 Btu/h × hp	CTI ATC-106
Centrifugal fan evaporative condensers	All	R-507A Test Fluid 165°F entering gas temperature 105°F condensing temperature 75°F entering wb	≥ 135,000 Btu/h × hp	CTI ATC-106
Air-cooled condensers	All	125°F Condensing Temperature 190°F Entering Gas Temperature 15°F subcooling 95°F entering db	≥ 176,000 Btu/h × hp	AHRI 460

For SI: °C = [(°F) - 32]/1.8, L/s • kW = (gpm/hp)/(11.83), COP = (Btu/h • hp)/(2550.7),
db = dry bulb temperature, °F, wb = wet bulb temperature, °F.

a. The efficiencies and test procedures for both open- and closed-circuit cooling towers are not applicable to hybrid cooling towers that contain a combination of wet and dry heat exchange sections.

b. For purposes of this table, open circuit cooling tower performance is defined as the water flow rating of the tower at the thermal rating condition, divided by the fan nameplate-rated motor power.

c. For purposes of this table, closed-circuit cooling tower performance is defined as the water flow rating of the tower at the thermal rating condition, divided by the sum of the fan nameplate-rated motor power and the spray pump nameplate-rated motor power.

d. For purposes of this table, air-cooled condenser performance is defined as the heat rejected from the refrigerant divided by the fan nameplate-rated motor power.

e. Chapter 6 contains a complete specification of the referenced test procedure, including the referenced year version of the test procedure. The certification requirements do not apply to field-erected cooling towers.

f. Where a certification program exists for a covered product and it includes provisions for verification and challenge of equipment efficiency ratings, then the product shall be listed in the certification program; or, where a certification program exists for a covered product, and it includes provisions for verification and challenge of equipment efficiency ratings, but the product is not listed in the existing certification program, the ratings shall be verified by an independent laboratory test report.

g. Cooling towers shall comply with the minimum efficiency listed in the table for that specific type of tower with the capacity effect of any project-specific accessories or options included in the capacity of the cooling tower

h. For purposes of this table, evaporative condenser performance is defined as the heat rejected at the specified rating condition in the table divided by the sum of the fan motor nameplate power and the integral spray pump nameplate power

i. Requirements for evaporative condensers are listed with ammonia (R-717) and R-507A as test fluids in the table. Evaporative condensers intended for use with halocarbon refrigerants other than R-507A shall meet the minimum efficiency requirements listed in this table with R-507A as the test fluid.

TABLE C403.3.2(9)
MINIMUM EFFICIENCY AIR CONDITIONERS AND CONDENSING UNITS SERVING COMPUTER ROOMS

EQUIPMENT TYPE	NET SENSIBLE COOLING CAPACITY[a]	MINIMUM SCOP-127[b] EFFICIENCY DOWNFLOW UNITS / UPFLOW UNITS	TEST PROCEDURE
Air conditioners, air cooled	< 65,000 Btu/h	2.20 / 2.09	ANSI/ASHRAE 127
	≥ 65,000 Btu/h and < 240,000 Btu/h	2.10 / 1.99	
	≥ 240,000 Btu/h	1.90 / 1.79	
Air conditioners, water cooled	< 65,000 Btu/h	2.60 / 2.49	
	≥ 65,000 Btu/h and < 240,000 Btu/h	2.50 / 2.39	
	≥ 240,000 Btu/h	2.40 /2.29	
Air conditioners, water cooled with fluid economizer	< 65,000 Btu/h	2.55 /2.44	
	≥ 65,000 Btu/h and < 240,000 Btu/h	2.45 / 2.34	
	≥ 240,000 Btu/h	2.35 / 2.24	
Air conditioners, glycol cooled (rated at 40% propylene glycol)	< 65,000 Btu/h	2.50 / 2.39	
	≥ 65,000 Btu/h and < 240,000 Btu/h	2.15 / 2.04	
	≥ 240,000 Btu/h	2.10 / 1.99	
Air conditioners, glycol cooled (rated at 40% propylene glycol) with fluid economizer	< 65,000 Btu/h	2.45 / 2.34	
	≥ 65,000 Btu/h and < 240,000 Btu/h	2.10 / 1.99	
	≥ 240,000 Btu/h	2.05 / 1.94	

For SI: 1 British thermal unit per hour = 0.2931 W.

a. Net sensible cooling capacity: the total gross cooling capacity less the latent cooling less the energy to the air movement system. (Total Gross – latent – Fan Power).

b. Sensible coefficient of performance (SCOP-127): a ratio calculated by dividing the net sensible cooling capacity in watts by the total power input in watts (excluding reheaters and humidifiers) at conditions defined in ASHRAE Standard 127. The net sensible cooling capacity is the gross sensible capacity minus the energy dissipated into the cooled space by the fan system.

TABLE C403.3.2(10)
HEAT TRANSFER EQUIPMENT

EQUIPMENT TYPE	SUBCATEGORY	MINIMUM EFFICIENCY	TEST PROCEDURE[a]
Liquid-to-liquid heat exchangers	Plate type	NR	AHRI 400

NR = No Requirement.

a. Chapter 6 contains a complete specification of the referenced test procedure, including the referenced year version of the test procedure.

C403.4.1.2 Deadband (Mandatory). Where used to control both heating and cooling, *zone* thermostatic controls shall be configured to provide a temperature range or deadband of not less than 5°F (2.8°C) within which the supply of heating and cooling energy to the *zone* is shut off or reduced to a minimum.

Exceptions:

1. Thermostats requiring manual changeover between heating and cooling modes.
2. Occupancies or applications requiring precision in indoor temperature control as *approved* by the *code official*.

C403.4.1.3 Setpoint overlap restriction (Mandatory). Where a *zone* has a separate heating and a separate cooling thermostatic control located within the *zone*, a limit switch, mechanical stop or direct digital control system with software programming shall be configured to prevent the heating setpoint from exceeding the cooling setpoint and to maintain a deadband in accordance with Section C403.4.1.2.

C403.4.1.4 Heated or cooled vestibules (Mandatory). The heating system for heated vestibules and air curtains with integral heating shall be provided with controls configured to shut off the source of heating when the outdoor air temperature is greater than 45°F (7°C). Vestibule heating and cooling systems shall be controlled by a thermostat located in the vestibule configured to limit heating to a temperature not greater than 60°F (16°C) and cooling to a temperature not less than 85°F (29°C).

Exception: Control of heating or cooling provided by site-recovered energy or transfer air that would otherwise be exhausted.

C403.4.1.5 Hot water boiler outdoor temperature setback control (Mandatory). Hot water boilers that supply heat to the building through one- or two-pipe heating systems shall have an outdoor setback control that lowers the boiler water temperature based on the outdoor temperature.

C403.4.2 Off-hour controls (Mandatory). Each *zone* shall be provided with thermostatic setback controls that are controlled by either an automatic time clock or programmable control system.

Exceptions:

1. *Zones* that will be operated continuously.
2. *Zones* with a full HVAC load demand not exceeding 6,800 Btu/h (2 kW) and having a manual shutoff switch located with *ready access*.

C403.4.2.1 Thermostatic setback (Mandatory). Thermostatic setback controls shall be configured to set back or temporarily operate the system to maintain zone temperatures down to 55°F (13°C) or up to 85°F (29°C).

C403.4.2.2 Automatic setback and shutdown (Mandatory). Automatic time clock or programmable controls shall be capable of starting and stopping the system for seven different daily schedules per week and retaining their programming and time setting during a loss of power for not fewer than 10 hours. Additionally, the controls shall have a manual override that allows temporary operation of the system for up to 2 hours; a manually operated timer configured to operate the system for up to 2 hours; or an occupancy sensor.

C403.4.2.3 Automatic start (Mandatory). Automatic start controls shall be provided for each HVAC system. The controls shall be configured to automatically adjust the daily start time of the HVAC system in order to bring each space to the desired occupied temperature immediately prior to scheduled occupancy.

C403.4.3 Hydronic systems controls. The heating of fluids that have been previously mechanically cooled and the cooling of fluids that have been previously mechanically heated shall be limited in accordance with Sections C403.4.3.1 through C403.4.3.3. Hydronic heating systems comprised of multiple-packaged boilers and designed to deliver conditioned water or steam into a common distribution system shall include automatic controls configured to sequence operation of the boilers. Hydronic heating systems composed of a single boiler and greater than 500,000 Btu/h (146.5 kW) input design capacity shall include either a multistaged or modulating burner.

C403.4.3.1 Three-pipe system. Hydronic systems that use a common return system for both hot water and chilled water are prohibited.

C403.4.3.2 Two-pipe changeover system. Systems that use a common distribution system to supply both heated and chilled water shall be designed to allow a deadband between changeover from one mode to the other of not less than 15°F (8.3°C) outside air temperatures; be designed to and provided with controls that will allow operation in one mode for not less than 4 hours before changing over to the other mode; and be provided with controls that allow heating and cooling supply temperatures at the changeover point to be not more than 30°F (16.7°C) apart.

C403.4.3.3 Hydronic (water loop) heat pump systems. Hydronic heat pump systems shall comply with Sections C403.4.3.3.1 through C403.4.3.3.3.

C403.4.3.3.1 Temperature deadband. Hydronic heat pumps connected to a common heat pump water loop with central devices for heat rejection and heat addition shall have controls that are configured to provide a heat pump water supply temperature deadband of not less than 20°F (11°C) between initiation of heat rejection and heat addition by the central devices.

Exception: Where a system loop temperature optimization controller is installed and can determine the most efficient operating temperature based on real-time conditions of demand and capacity, deadbands of less than 20°F (11°C) shall be permitted.

C403.4.3.3.2 Heat rejection. The following shall apply to hydronic water loop heat pump systems in Climate Zones 3 through 8:

1. Where a closed-circuit cooling tower is used directly in the heat pump loop, either an automatic valve shall be installed to bypass the flow of water around the closed-circuit cooling tower, except for any flow necessary for freeze protection, or low-leakage positive-closure dampers shall be provided.
2. Where an open-circuit cooling tower is used directly in the heat pump loop, an automatic valve shall be installed to bypass all heat pump water flow around the open-circuit cooling tower.
3. Where an open-circuit cooling tower is used in conjunction with a separate heat exchanger to isolate the open-circuit cooling tower from the heat pump loop, heat loss shall be controlled by shutting down the circulation pump on the cooling tower loop.

Exception: Where it can be demonstrated that a heat pump system will be required to reject heat throughout the year.

C403.4.3.3.3 Two-position valve. Each hydronic heat pump on the hydronic system having a total pump system power exceeding 10 hp (7.5 kW) shall have a two-position valve.

C403.4.4 Part-load controls. Hydronic systems greater than or equal to 300,000 Btu/h (87.9 kW) in design output capacity supplying heated or chilled water to comfort conditioning systems shall include controls that are configured to do all of the following:

1. Automatically reset the supply-water temperatures in response to varying building heating and cooling demand using coil valve position, zone-return water temperature, building-return water temperature or outside air temperature. The temperature shall be reset by not less than 25 percent of the design supply-to-return water temperature difference.

2. Automatically vary fluid flow for hydronic systems with a combined pump motor capacity of 2 hp (1.5 kW) or larger with three or more control valves or other devices by reducing the system design flow rate by not less than 50 percent or the maximum reduction allowed by the equipment manufacturer for proper operation of equipment by valves that modulate or step open and close, or pumps that modulate or turn on and off as a function of load.
3. Automatically vary pump flow on heating-water systems, chilled-water systems and heat rejection loops serving water-cooled unitary air conditioners as follows:
   3.1. Where pumps operate continuously or operate based on a time schedule, pumps with nominal output motor power of 2 hp or more shall have a variable speed drive.
   3.2. Where pumps have automatic direct digital control configured to operate pumps only when zone heating or cooling is required, a variable speed drive shall be provided for pumps with motors having the same or greater nominal output power indicated in Table C403.4.4 based on the climate zone and system served.
4. Where a variable speed drive is required by Item 3 of this Section, pump motor power input shall be not more than 30 percent of design wattage at 50 percent of the design water flow. Pump flow shall be controlled to maintain one control valve nearly wide open or to satisfy the minimum differential pressure.

Exceptions:

1. Supply-water temperature reset is not required for chilled-water systems supplied by off-site district chilled water or chilled water from ice storage systems.
2. Variable pump flow is not required on dedicated coil circulation pumps where needed for freeze protection.
3. Variable pump flow is not required on dedicated equipment circulation pumps where configured in primary/secondary design to provide the minimum flow requirements of the equipment manufacturer for proper operation of equipment.
4. Variable speed drives are not required on heating water pumps where more than 50 percent of annual heat is generated by an electric boiler.

C403.4.5 Pump isolation. Chilled water plants including more than one chiller shall be capable of and configured to reduce flow automatically through the chiller plant when a chiller is shut down. Chillers piped in series for the purpose of increased temperature differential shall be considered as one chiller.

Boiler systems including more than one boiler shall be capable of and configured to reduce flow automatically through the boiler system when a boiler is shut down.

C403.5 Economizers (Prescriptive). Economizers shall comply with Sections C403.5.1 through C403.5.5.

An air or water economizer shall be provided for the following cooling systems:

1. Chilled water systems with a total cooling capacity, less cooling capacity provided with air economizers, as specified in Table C403.5(1).
2. Individual fan systems with cooling capacity greater than or equal to 54,000 Btu/h (15.8 kW) in buildings having other than a *Group R* occupancy,

   The total supply capacity of all fan cooling units not provided with economizers shall not exceed 20 percent of the total supply capacity of all fan cooling units in the building or 300,000 Btu/h (88 kW), whichever is greater.
3. Individual fan systems with cooling capacity greater than or equal to 270,000 Btu/h (79.1 kW) in buildings having a *Group R* occupancy.

   The total supply capacity of all fan cooling units not provided with economizers shall not exceed 20 percent of the total supply capacity of all fan cooling units in

TABLE C403.4.4
VARIABLE SPEED DRIVE (VSD) REQUIREMENTS FOR DEMAND-CONTROLLED PUMPS

CHILLED WATER AND HEAT REJECTION LOOP PUMPS IN THESE CLIMATE ZONES	HEATING WATER PUMPS IN THESE CLIMATE ZONES	VSD REQUIRED FOR MOTORS WITH RATED OUTPUT OF:
1A, 1B, 2B	—	≥ 2 hp
2A, 3B	—	≥ 3 hp
3A, 3C, 4A, 4B	7, 8	≥ 5 hp
4C, 5A, 5B, 5C, 6A, 6B	3C, 5A, 5C, 6A, 6B	≥ 7.5 hp
—	4A, 4C, 5B	≥ 10 hp
7, 8	4B	≥ 15 hp
—	2A, 2B, 3A, 3B	≥ 25 hp
—	1B	≥ 100 hp
—	1A	≥ 200 hp

the building or 1,500,000 Btu/h (440 kW), whichever is greater.

Exceptions: Economizers are not required for the following systems.

1. Individual fan systems not served by chilled water for buildings located in *Climate Zones* 1A and 1B.
2. Where more than 25 percent of the air designed to be supplied by the system is to spaces that are designed to be humidified above 35°F (1.7°C) dew-point temperature to satisfy process needs.
3. Systems expected to operate less than 20 hours per week.
4. Systems serving supermarket areas with open refrigerated casework.
5. Where the cooling efficiency is greater than or equal to the efficiency requirements in Table C403.5(2).
6. Systems that include a heat recovery system in accordance with Section C403.9.5.

TABLE C403.5(2)
EQUIPMENT EFFICIENCY PERFORMANCE EXCEPTION FOR ECONOMIZERS

CLIMATE ZONES	COOLING EQUIPMENT PERFORMANCE IMPROVEMENT (EER OR IPLV)
2A, 2B	10% efficiency improvement
3A, 3B	15% efficiency improvement
4A, 4B	20% efficiency improvement

C403.5.1 Integrated economizer control. Economizer systems shall be integrated with the mechanical cooling system and be configured to provide partial cooling even where additional mechanical cooling is required to provide the remainder of the cooling load. Controls shall not be capable of creating a false load in the mechanical cooling systems by limiting or disabling the economizer or any other means, such as hot gas bypass, except at the lowest stage of mechanical cooling.

Units that include an air economizer shall comply with the following:

1. Unit controls shall have the mechanical cooling capacity control interlocked with the air economizer controls such that the outdoor air damper is at the 100-percent open position when mechanical cooling is on and the outdoor air damper does not begin to close to prevent coil freezing due to minimum compressor run time until the leaving air temperature is less than 45°F (7°C).
2. Direct expansion (DX) units that control 75,000 Btu/h (22 kW) or greater of rated capacity of the capacity of the mechanical cooling directly based on occupied space temperature shall have not fewer than two stages of mechanical cooling capacity.
3. Other DX units, including those that control space temperature by modulating the airflow to the space, shall be in accordance with Table C403.5.1.

C403.5.2 Economizer heating system impact. HVAC system design and economizer controls shall be such that economizer operation does not increase building heating energy use during normal operation.

Exception: Economizers on variable air volume (VAV) systems that cause zone level heating to increase because of a reduction in supply air temperature.

C403.5.3 Air economizers. Where economizers are required by Section C403.5, air economizers shall comply with Sections C403.5.3.1 through C403.5.3.5.

C403.5.3.1 Design capacity. Air economizer systems shall be configured to modulate *outdoor air* and return

TABLE C403.5(1)
MINIMUM CHILLED-WATER SYSTEM COOLING CAPACITY FOR DETERMINING ECONOMIZER COOLING REQUIREMENTS

CLIMATE ZONES (COOLING)	TOTAL CHILLED-WATER SYSTEM CAPACITY LESS CAPACITY OF COOLING UNITS WITH AIR ECONOMIZERS	
	Local Water-cooled Chilled-water Systems	Air-cooled Chilled-water Systems or District Chilled-Water Systems
1A	Economizer not required	Economizer not required
1B, 2A, 2B	960,000 Btu/h	1,250,000 Btu/h
3A, 3B, 3C, 4A, 4B, 4C	720,000 Btu/h	940,000 Btu/h
5A, 5B, 5C, 6A, 6B, 7, 8	1,320,000 Btu/h	1,720,000 Btu/h

For SI: 1 British thermal unit per hour = 0.2931 W.

TABLE C403.5.1
DX COOLING STAGE REQUIREMENTS FOR MODULATING AIRFLOW UNITS

RATING CAPACITY	MINIMUM NUMBER OF MECHANICAL COOLING STAGES	MINIMUM COMPRESSOR DISPLACEMENT[a]
≥ 65,000 Btu/h and < 240,000 Btu/h	3 stages	≤ 35% of full load
≥ 240,000 Btu/h	4 stages	≤ 25% full load

For SI: 1 British thermal unit per hour = 0.2931 W.

a. For mechanical cooling stage control that does not use variable compressor displacement, the percent displacement shall be equivalent to the mechanical cooling capacity reduction evaluated at the full load rating conditions for the compressor.

air dampers to provide up to 100 percent of the design supply air quantity as *outdoor air* for cooling.

C403.5.3.2 Control signal. Economizer controls and dampers shall be configured to sequence the dampers with the mechanical cooling equipment and shall not be controlled by only mixed-air temperature.

Exception: The use of mixed-air temperature limit control shall be permitted for systems controlled from space temperature (such as single-*zone* systems).

C403.5.3.3 High-limit shutoff. Air economizers shall be configured to automatically reduce *outdoor air* intake to the design minimum *outdoor air* quantity when *outdoor air* intake will not reduce cooling energy usage. High-limit shutoff control types for specific climates shall be chosen from Table C403.5.3.3. High-limit shutoff control settings for these control types shall be those specified in Table C403.5.3.3.

C403.5.3.4 Relief of excess outdoor air. Systems shall be capable of relieving excess *outdoor air* during air economizer operation to prevent overpressurizing the building. The relief air outlet shall be located to avoid recirculation into the building.

C403.5.3.5 Economizer dampers. Return, exhaust/relief and outdoor air dampers used in economizers shall comply with Section C403.7.7.

C403.5.4 Water-side economizers. Where economizers are required by Section C403.5, water-side economizers shall comply with Sections C403.5.4.1 and C403.5.4.2.

C403.5.4.1 Design capacity. Water economizer systems shall be configured to cool supply air by indirect evaporation and providing up to 100 percent of the expected system cooling load at *outdoor air* temperatures of not greater than 50°F (10°C) dry bulb/45°F (7°C) wet bulb.

Exceptions:

1. Systems primarily serving computer rooms in which 100 percent of the expected system cooling load at 40°F (4°C) dry bulb/35°F (1.7°C) wet bulb is met with evaporative water economizers.
2. Systems primarily serving computer rooms with dry cooler water economizers that satisfy 100 percent of the expected system cooling load at 35°F (1.7°C) dry bulb.
3. Systems where dehumidification requirements cannot be met using outdoor air temperatures of 50°F (10°C) dry bulb/45°F (7°C) wet bulb and where 100 percent of the expected system cooling load at 45°F (7°C) dry bulb/40°F (4°C) wet bulb is met with evaporative water economizers.

C403.5.4.2 Maximum pressure drop. Precooling coils and water-to-water heat exchangers used as part of a water economizer system shall either have a water-side pressure drop of less than 15 feet (45 kPa) of water or a secondary loop shall be created so that the coil or heat exchanger pressure drop is not seen by the circulating pumps when the system is in the normal cooling (noneconomizer) mode.

C403.5.5 Economizer fault detection and diagnostics (Mandatory). Air-cooled unitary direct-expansion units listed in Tables C403.3.2(1) through C403.3.2(3) and variable refrigerant flow (VRF) units that are equipped with an

TABLE C403.5.3.3
HIGH-LIMIT SHUTOFF CONTROL SETTING FOR AIR ECONOMIZERS[b]

DEVICE TYPE	CLIMATE ZONE	REQUIRED HIGH LIMIT (ECONOMIZER OFF WHEN):	
		Equation	Description
Fixed dry bulb	1B, 2B, 3B, 3C, 4B, 4C, 5B, 5C, 6B, 7, 8	$T_{OA} > 75°F$	Outdoor air temperature exceeds 75°F
	5A, 6A	$T_{OA} > 70°F$	Outdoor air temperature exceeds 70°F
	1A, 2A, 3A, 4A	$T_{OA} > 65°F$	Outdoor air temperature exceeds 65°F
Differential dry bulb	1B, 2B, 3B, 3C, 4B, 4C, 5A, 5B, 5C, 6A, 6B, 7, 8	$T_{OA} > T_{RA}$	Outdoor air temperature exceeds return air temperature
Fixed enthalpy with fixed dry-bulb temperatures	All	$h_{OA} > 28$ Btu/lb[a] or $T_{OA} > 75°F$	Outdoor air enthalpy exceeds 28 Btu/lb of dry air[a] or Outdoor air temperature exceeds 75°F
Differential enthalpy with fixed dry-bulb temperature	All	$h_{OA} > h_{RA}$ or $T_{OA} > 75°F$	Outdoor air enthalpy exceeds return air enthalpy or Outdoor air temperature exceeds 75°F

For SI: 1 foot = 305 mm, °C = (°F - 32)/1.8, 1 Btu/lb = 2.33 kJ/kg.

a. At altitudes substantially different than sea level, the fixed enthalpy limit shall be set to the enthalpy value at 75°F and 50-percent relative humidity. As an example, at approximately 6,000 feet elevation, the fixed enthalpy limit is approximately 30.7 Btu/lb.

b. Devices with selectable setpoints shall be capable of being set to within 2°F and 2 Btu/lb of the setpoint listed.

economizer in accordance with Sections C403.5 through C403.5.4 shall include a fault detection and diagnostics system complying with the following:

1. The following temperature sensors shall be permanently installed to monitor system operation:
   1.1. Outside air.
   1.2. Supply air.
   1.3. Return air.
2. Temperature sensors shall have an accuracy of ±2°F (1.1°C) over the range of 40°F to 80°F (4°C to 26.7°C).
3. Refrigerant pressure sensors, where used, shall have an accuracy of ±3 percent of full scale.
4. The unit controller shall be configured to provide system status by indicating the following:
   4.1. Free cooling available.
   4.2. Economizer enabled.
   4.3. Compressor enabled.
   4.4. Heating enabled.
   4.5. Mixed air low limit cycle active.
   4.6. The current value of each sensor.
5. The unit controller shall be capable of manually initiating each operating mode so that the operation of compressors, economizers, fans and the heating system can be independently tested and verified.
6. The unit shall be configured to report faults to a fault management application available for *access* by day-to-day operating or service personnel, or annunciated locally on zone thermostats.
7. The fault detection and diagnostics system shall be configured to detect the following faults:
   7.1. Air temperature sensor failure/fault.
   7.2. Not economizing when the unit should be economizing.
   7.3. Economizing when the unit should not be economizing.
   7.4. Damper not modulating.
   7.5. Excess outdoor air.

C403.6 Requirements for mechanical systems serving multiple zones. Sections C403.6.1 through C403.6.9 shall apply to mechanical systems serving multiple zones.

C403.6.1 Variable air volume and multiple-zone systems. Supply air systems serving multiple zones shall be variable air volume (VAV) systems that have zone controls configured to reduce the volume of air that is reheated, recooled or mixed in each zone to one of the following:

1. Twenty percent of the zone design peak supply for systems with DDC and 30 percent for other systems.
2. Systems with DDC where all of the following apply:
   2.1. The airflow rate in the deadband between heating and cooling does not exceed 20 percent of the zone design peak supply rate or higher allowed rates under Items 3, 4 and 5 of this section.
   2.2. The first stage of heating modulates the zone supply air temperature setpoint up to a maximum setpoint while the airflow is maintained at the deadband flow rate.
   2.3. The second stage of heating modulates the airflow rate from the deadband flow rate up to the heating maximum flow rate that is less than 50 percent of the zone design peak supply rate.
3. The outdoor airflow rate required to meet the minimum ventilation requirements of Chapter 4 of the *International Mechanical Code*.
4. Any higher rate that can be demonstrated to reduce overall system annual energy use by offsetting reheat/recool energy losses through a reduction in outdoor air intake for the system as approved by the code official.
5. The airflow rate required to comply with applicable codes or accreditation standards such as pressure relationships or minimum air change rates.

Exception: The following individual zones or entire air distribution systems are exempted from the requirement for VAV control:

1. *Zones* or supply air systems where not less than 75 percent of the energy for reheating or for providing warm air in mixing systems is provided from a site-recovered, including condenser heat, or site-solar energy source.
2. Systems that prevent reheating, recooling, mixing or simultaneous supply of air that has been previously cooled, either mechanically or through the use of economizer systems, and air that has been previously mechanically heated.

C403.6.2 Single-duct VAV systems, terminal devices. Single-duct VAV systems shall use terminal devices capable of and configured to reduce the supply of primary supply air before reheating or recooling takes place.

C403.6.3 Dual-duct and mixing VAV systems, terminal devices. Systems that have one warm air duct and one cool air duct shall use terminal devices that are configured to reduce the flow from one duct to a minimum before mixing of air from the other duct takes place.

C403.6.4 Single-fan dual-duct and mixing VAV systems, economizers. Individual dual-duct or mixing heating and cooling systems with a single fan and with total capacities greater than 90,000 Btu/h [(26.4 kW) 7.5 tons] shall not be equipped with air economizers.

C403.6.5 Supply-air temperature reset controls. Multiple-*zone* HVAC systems shall include controls that automatically reset the supply-air temperature in response to representative building loads, or to outdoor air temperature. The controls shall be configured to reset the supply air temperature not less than 25 percent of the difference between the design supply-air temperature and the design room air temperature.

Exceptions:

1. Systems that prevent reheating, recooling or mixing of heated and cooled supply air.
2. Seventy-five percent of the energy for reheating is from site-recovered or site-solar energy sources.
3. *Zones* with peak supply air quantities of 300 cfm (142 L/s) or less.

C403.6.6 Multiple-zone VAV system ventilation optimization control. Multiple-zone VAV systems with direct digital control of individual zone boxes reporting to a central control panel shall have automatic controls configured to reduce outdoor air intake flow below design rates in response to changes in system *ventilation* efficiency (E_v) as defined by the *International Mechanical Code*.

Exceptions:

1. VAV systems with zonal transfer fans that recirculate air from other zones without directly mixing it with outdoor air, dual-duct dual-fan VAV systems, and VAV systems with fan-powered terminal units.
2. Systems where total design exhaust airflow is more than 70 percent of total design outdoor air intake flow requirements.

C403.6.7 Parallel-flow fan-powered VAV air terminal control. Parallel-flow fan-powered VAV air terminals shall have automatic controls configured to:

1. Turn off the terminal fan except when space heating is required or where required for ventilation.
2. Turn on the terminal fan as the first stage of heating before the heating coil is activated.
3. During heating for warmup or setback temperature control, either:
   - 3.1. Operate the terminal fan and heating coil without primary air.
   - 3.2. Reverse the terminal damper logic and provide heating from the central air handler by primary air.

C403.6.8 Setpoints for direct digital control. For systems with direct digital control of individual zones reporting to the central control panel, the static pressure setpoint shall be reset based on the *zone* requiring the most pressure. In such case, the setpoint is reset lower until one *zone* damper is nearly wide open. The direct digital controls shall be capable of monitoring zone damper positions or shall have an alternative method of indicating the need for static pressure that is configured to provide all of the following:

1. Automatic detection of any *zone* that excessively drives the reset logic.
2. Generation of an alarm to the system operational location.
3. Allowance for an operator to readily remove one or more *zones* from the reset algorithm.

C403.6.9 Static pressure sensor location. Static pressure sensors used to control VAV fans shall be located such that the controller setpoint is not greater than 1.2 inches w.c. (299 Pa). Where this results in one or more sensors being located downstream of major duct splits, not less than one sensor shall be located on each major branch to ensure that static pressure can be maintained in each branch.

C403.7 Ventilation and exhaust systems (Mandatory). In addition to other requirements of Section C403 applicable to the provision of ventilation air or the exhaust of air, ventilation and exhaust systems shall be in accordance with Sections C403.7.1 through C403.7.7.

C403.7.1 Demand control ventilation (Mandatory). Demand control ventilation (DCV) shall be provided for spaces larger than 500 square feet (46.5 m^2) and with an average occupant load of 25 people or greater per 1,000 square feet (93 m^2) of floor area, as established in Table 403.3.1.1 of the *International Mechanical Code*, and served by systems with one or more of the following:

1. An air-side economizer.
2. Automatic modulating control of the outdoor air damper.
3. A design outdoor airflow greater than 3,000 cfm (1416 L/s).

Exceptions:

1. Systems with energy recovery complying with Section C403.7.4.
2. Multiple-zone systems without direct digital control of individual zones communicating with a central control panel.
3. Systems with a design outdoor airflow less than 1,200 cfm (566 L/s).
4. Spaces where the supply airflow rate minus any makeup or outgoing transfer air requirement is less than 1,200 cfm (566 L/s).
5. Ventilation provided only for process loads.

C403.7.2 Enclosed parking garage ventilation controls (Mandatory). Enclosed parking garages used for storing or handling automobiles operating under their own power shall employ contamination-sensing devices and automatic controls configured to stage fans or modulate fan average airflow rates to 50 percent or less of design capacity, or intermittently operate fans less than 20 percent of the occupied time or as required to maintain acceptable contaminant levels in accordance with *International Mechanical Code* provisions. Failure of contamination-sensing devices shall cause the exhaust fans to operate continuously at design airflow.

Exceptions:

1. Garages with a total exhaust capacity less than 22,500 cfm (10 620 L/s) with ventilation systems that do not utilize heating or mechanical cooling.

2. Garages that have a garage area to ventilation system motor nameplate power ratio that exceeds 1125 cfm/hp (710 L/s/kW) and do not utilize heating or mechanical cooling.

C403.7.3 Ventilation air heating control (Mandatory). Units that provide ventilation air to multiple zones and operate in conjunction with zone heating and cooling systems shall not use heating or heat recovery to warm supply air to a temperature greater than 60°F (16°C) when representative building loads or outdoor air temperatures indicate that the majority of zones require cooling.

C403.7.4 Energy recovery ventilation systems (Mandatory). Where the supply airflow rate of a fan system exceeds the values specified in Tables C403.7.4(1) and C403.7.4(2), the system shall include an energy recovery system. The energy recovery system shall be configured to provide a change in the enthalpy of the outdoor air supply of not less than 50 percent of the difference between the outdoor air and return air enthalpies, at design conditions. Where an air economizer is required, the energy recovery system shall include a bypass or controls that permit operation of the economizer as required by Section C403.5.

Exception: An energy recovery ventilation system shall not be required in any of the following conditions:

1. Where energy recovery systems are prohibited by the *International Mechanical Code*.
2. Laboratory fume hood systems that include not fewer than one of the following features:
   - 2.1. Variable-air-volume hood exhaust and room supply systems configured to reduce exhaust and makeup air volume to 50 percent or less of design values.
   - 2.2. Direct makeup (auxiliary) air supply equal to or greater than 75 percent of the exhaust rate, heated not warmer than 2°F (1.1°C) above room setpoint, cooled to not cooler than 3°F (1.7°C) below room setpoint, with no humidification added, and no simultaneous heating and cooling used for dehumidification control.
3. Systems serving spaces that are heated to less than 60°F (15.5°C) and that are not cooled.
4. Where more than 60 percent of the outdoor heating energy is provided from site-recovered or site-solar energy.
5. Heating energy recovery in *Climate Zones* 1 and 2.
6. Cooling energy recovery in *Climate Zones* 3C, 4C, 5B, 5C, 6B, 7 and 8.

TABLE C403.7.4(1)
ENERGY RECOVERY REQUIREMENT
(Ventilation systems operating less than 8,000 hours per year)

CLIMATE ZONE	PERCENT (%) OUTDOOR AIR AT FULL DESIGN AIRFLOW RATE							
	≥ 10% and < 20%	≥ 20% and < 30%	≥ 30% and < 40%	≥ 40% and < 50%	≥ 50% and < 60%	≥ 60% and < 70%	≥ 70% and < 80%	≥ 80%
	DESIGN SUPPLY FAN AIRFLOW RATE (cfm)							
3B, 3C, 4B, 4C, 5B	NR	NR	NR	NR	NR	NR	NR	NR
1B, 2B, 5C	NR	NR	NR	NR	≥ 26,000	≥ 12,000	≥ 5,000	≥ 4,000
6B	≥ 28,000	≥ 26,5000	≥ 11,000	≥ 5,500	≥ 4,500	≥ 3,500	≥ 2,500	≥ 1,500
1A, 2A, 3A, 4A, 5A, 6A	≥ 26,000	≥ 16,000	≥ 5,500	≥ 4,500	≥ 3,500	≥ 2,000	≥ 1,000	> 120
7, 8	≥ 4,500	≥ 4,000	≥ 2,500	≥ 1,000	>140	> 120	> 100	> 80

For SI: 1 cfm = 0.4719 L/s.
NR = Not Required.

TABLE C403.7.4(2)
ENERGY RECOVERY REQUIREMENT
(Ventilation systems operating not less than 8,000 hours per year)

CLIMATE ZONE	PERCENT (%) OUTDOOR AIR AT FULL DESIGN AIRFLOW RATE							
	≥ 10% and < 20%	≥ 20% and < 30%	≥ 30% and < 40%	≥ 40% and < 50%	≥ 50% and < 60%	≥ 60% and < 70%	≥ 70% and < 80%	≥ 80%
	Design Supply Fan Airflow Rate (cfm)							
3C	NR	NR	NR	NR	NR	NR	NR	NR
1B, 2B, 3B, 4C, 5C	NR	≥ 19,500	≥ 9,000	≥ 5,000	≥ 4,000	≥ 3,000	≥ 1,500	≥ 120
1A, 2A, 3A, 4B, 5B	≥ 2,500	≥ 2,000	≥ 1,000	≥ 500	≥ 140	≥ 120	≥ 100	≥ 80
4A, 5A, 6A, 6B, 7, 8	≥ 200	≥ 130	≥ 100	≥ 80	≥ 70	≥ 60	≥ 50	≥ 40

For SI: 1 cfm = 0.4719 L/s.
NR = Not Required.

7. Systems requiring dehumidification that employ energy recovery in series with the cooling coil.
8. Where the largest source of air exhausted at a single location at the building exterior is less than 75 percent of the design *outdoor air* flow rate.
9. Systems expected to operate less than 20 hours per week at the *outdoor air* percentage covered by Table C403.7.4(1).
10. Systems exhausting toxic, flammable, paint or corrosive fumes or dust.
11. Commercial kitchen hoods used for collecting and removing grease vapors and smoke.

C403.7.5 Kitchen exhaust systems (Mandatory). Replacement air introduced directly into the exhaust hood cavity shall not be greater than 10 percent of the hood exhaust airflow rate. Conditioned supply air delivered to any space shall not exceed the greater of the following:

1. The ventilation rate required to meet the space heating or cooling load.
2. The hood exhaust flow minus the available transfer air from adjacent space where available transfer air is considered to be that portion of outdoor ventilation air not required to satisfy other exhaust needs, such as restrooms, and not required to maintain pressurization of adjacent spaces.

Where total kitchen hood exhaust airflow rate is greater than 5,000 cfm (2360 L/s), each hood shall be a factory-built commercial exhaust hood listed by a nationally recognized testing laboratory in compliance with UL 710. Each hood shall have a maximum exhaust rate as specified in Table C403.7.5 and shall comply with one of the following:

1. Not less than 50 percent of all replacement air shall be transfer air that would otherwise be exhausted.
2. Demand ventilation systems on not less than 75 percent of the exhaust air that are configured to provide not less than a 50-percent reduction in exhaust and replacement air system airflow rates, including controls necessary to modulate airflow in response to appliance operation and to maintain full capture and containment of smoke, effluent and combustion products during cooking and idle.
3. Listed energy recovery devices with a sensible heat recovery effectiveness of not less than 40 percent on not less than 50 percent of the total exhaust airflow.

Where a single hood, or hood section, is installed over appliances with different duty ratings, the maximum allowable flow rate for the hood or hood section shall be based on the requirements for the highest appliance duty rating under the hood or hood section.

Exception: Where not less than 75 percent of all the replacement air is transfer air that would otherwise be exhausted.

C403.7.6 Automatic control of HVAC systems serving guestrooms (Mandatory). In *Group R*-1 buildings containing more than 50 guestrooms, each guestroom shall be provided with controls complying with the provisions of Sections C403.7.6.1 and C403.7.6.2. Card key controls comply with these requirements.

C403.7.6.1 Temperature setpoint controls. Controls shall be provided on each HVAC system that are capable of and configured to automatically raise the cooling setpoint and lower the heating setpoint by not less than 4°F (2°C) from the occupant setpoint within 30 minutes after the occupants have left the guestroom. The controls shall be capable of and configured to automatically raise the cooling setpoint to not lower than 80°F (27°C) and lower the heating setpoint to not higher than 60°F (16°C) when the guestroom is unrented or has not been continuously occupied for more than 16 hours or a *networked guestroom control system* indicates that the guestroom is unrented and the guestroom is unoccupied for more than 30 minutes. A *networked guestroom control system* that is capable of returning the thermostat setpoints to default occupied setpoints 60 minutes prior to the time a guestroom is scheduled to be occupied is not precluded by this section. Cooling that is capable of limiting relative humidity with a setpoint not lower than 65-percent relative humidity during unoccupied periods is not precluded by this section.

C403.7.6.2 Ventilation controls. Controls shall be provided on each HVAC system that are capable of and configured to automatically turn off the ventilation and

TABLE C403.7.5
MAXIMUM NET EXHAUST FLOW RATE,
CFM PER LINEAR FOOT OF HOOD LENGTH

TYPE OF HOOD	LIGHT-DUTY EQUIPMENT	MEDIUM-DUTY EQUIPMENT	HEAVY-DUTY EQUIPMENT	EXTRA-HEAVY-DUTY EQUIPMENT
Wall-mounted canopy	140	210	280	385
Single island	280	350	420	490
Double island (per side)	175	210	280	385
Eyebrow	175	175	NA	NA
Backshelf/Pass-over	210	210	280	NA

For SI:1 cfm = 0.4719 L/s; 1 foot = 305 mm.
NA = Not Allowed.

exhaust fans within 30 minutes of the occupants leaving the guestroom, or *isolation devices* shall be provided to each guestroom that are capable of automatically shutting off the supply of outdoor air to and exhaust air from the guestroom.

Exception: Guestroom ventilation systems are not precluded from having an automatic daily pre-occupancy purge cycle that provides daily outdoor air ventilation during unrented periods at the design ventilation rate for 60 minutes, or at a rate and duration equivalent to one air change.

C403.7.7 Shutoff dampers (Mandatory). Outdoor air intake and exhaust openings and stairway and shaft vents shall be provided with Class I motorized dampers. The dampers shall have an air leakage rate not greater than 4 cfm/ft^2 (20.3 L/s • m^2) of damper surface area at 1.0 inch water gauge (249 Pa) and shall be labeled by an approved agency when tested in accordance with AMCA 500D for such purpose.

Outdoor air intake and exhaust dampers shall be installed with automatic controls configured to close when the systems or spaces served are not in use or during unoccupied period warm-up and setback operation, unless the systems served require outdoor or exhaust air in accordance with the *International Mechanical Code* or the dampers are opened to provide intentional economizer cooling.

Stairway and shaft vent dampers shall be installed with automatic controls configured to open upon the activation of any fire alarm initiating device of the building's fire alarm system or the interruption of power to the damper.

Exception: Nonmotorized gravity dampers shall be an alternative to motorized dampers for exhaust and relief openings as follows:

1. In buildings less than three stories in height above grade plane.
2. In buildings of any height located in *Climate Zones* 1, 2 or 3.
3. Where the design exhaust capacity is not greater than 300 cfm (142 L/s).

Nonmotorized gravity dampers shall have an air leakage rate not greater than 20 cfm/ft^2 (101.6 L/s • m^2) where not less than 24 inches (610 mm) in either dimension and 40 cfm/ft^2 (203.2 L/s • m^2) where less than 24 inches (610 mm) in either dimension. The rate of air leakage shall be determined at 1.0 inch water gauge (249 Pa) when tested in accordance with AMCA 500D for such purpose. The dampers shall be labeled by an approved agency.

C403.8 Fans and fan controls. Fans in HVAC systems shall comply with Sections C403.8.1 through C403.8.5.1.

C403.8.1 Allowable fan horsepower (Mandatory). Each HVAC system having a total fan system motor nameplate horsepower exceeding 5 hp (3.7 kW) at fan system design conditions shall not exceed the allowable *fan system motor nameplate hp* (Option 1) or *fan system bhp* (Option 2) shown in Table C403.8.1(1). This includes supply fans, exhaust fans, return/relief fans, and fan-powered terminal units associated with systems providing heating or cooling capability. Single-zone variable air volume systems shall comply with the constant volume fan power limitation.

Exceptions:

1. Hospital, vivarium and laboratory systems that utilize flow control devices on exhaust or return to maintain space pressure relationships necessary for occupant health and safety or environmental control shall be permitted to use variable volume fan power limitation.
2. Individual exhaust fans with motor nameplate horsepower of 1 hp (0.746 kW) or less are exempt from the allowable fan horsepower requirement.

C403.8.2 Motor nameplate horsepower (Mandatory). For each fan, the fan brake horsepower shall be indicated on the construction documents and the selected motor shall be not larger than the first available motor size greater than the following:

1. For fans less than 6 bhp (4413 W), 1.5 times the fan brake horsepower.
2. For fans 6 bhp (4413 W) and larger, 1.3 times the fan brake horsepower.
3. Systems complying with Section C403.8.1 *fan system motor nameplate hp* (Option 1).

Exception: Fans with motor nameplate horsepower less than 1 hp (746 W) are exempt from this section.

TABLE C403.8.1(1)
FAN POWER LIMITATION

	LIMIT	CONSTANT VOLUME	VARIABLE VOLUME
Option 1: Fan system motor nameplate hp	Allowable nameplate motor hp	$hp \le CFM_S \times 0.0011$	$hp \le CFM_S \times 0.0015$
Option 2: Fan system bhp	Allowable fan system bhp	$bhp \le CFM_S \times 0.00094 + A$	$bhp \le CFM_S \times 0.0013 + A$

For SI: 1 bhp = 735.5 W, 1 hp = 745.5 W, 1 cfm = 0.4719 L/s.

where:

CFM_S = The maximum design supply airflow rate to conditioned spaces served by the system in cubic feet per minute.

hp = The maximum combined motor nameplate horsepower.

bhp = The maximum combined fan brake horsepower.

A = Sum of $[PD \times CFM_D / 4131]$.

where:

PD = Each applicable pressure drop adjustment from Table C403.8.1(2) in. w.c.

CFM_D = The design airflow through each applicable device from Table C403.8.1(2) in cubic feet per minute.

**TABLE C403.8.1(2)
FAN POWER LIMITATION PRESSURE DROP ADJUSTMENT**

DEVICE	ADJUSTMENT
	Credits
Return air or exhaust systems required by code or accreditation standards to be fully ducted, or systems required to maintain air pressure differentials between adjacent rooms	0.5 inch w.c. (2.15 inches w.c. for laboratory and vivarium systems)
Return and exhaust airflow control devices	0.5 inch w.c.
Exhaust filters, scrubbers or other exhaust treatment	The pressure drop of device calculated at fan system design condition
Particulate filtration credit: MERV 9 thru 12	0.5 inch w.c.
Particulate filtration credit: MERV 13 thru 15	0.9 inch w.c.
Particulate filtration credit: MERV 16 and greater and electronically enhanced filters	Pressure drop calculated at 2x clean filter pressure drop at fan system design condition.
Carbon and other gas-phase air cleaners	Clean filter pressure drop at fan system design condition.
Biosafety cabinet	Pressure drop of device at fan system design condition.
Energy recovery device, other than coil runaround loop	For each airstream, (2.2 × energy recovery effectiveness – 0.5) inch w.c.
Coil runaround loop	0.6 inch w.c. for each airstream.
Evaporative humidifier/cooler in series with another cooling coil	Pressure drop of device at fan system design conditions.
Sound attenuation section (fans serving spaces with design background noise goals below NC35)	0.15 inch w.c.
Exhaust system serving fume hoods	0.35 inch w.c.
Laboratory and vivarium exhaust systems in high-rise buildings	0.25 inch w.c./100 feet of vertical duct exceeding 75 feet.
	Deductions
Systems without central cooling device	- 0.6 inch w.c.
Systems without central heating device	- 0.3 inch w.c.
Systems with central electric resistance heat	- 0.2 inch w.c.

For SI: 1 inch w.c. = 249 Pa, 1 inch = 25.4 mm.
w.c. = water column, NC = Noise criterion.

C403.8.3 Fan efficiency (Mandatory). Fans shall have a fan efficiency grade (FEG) of not less than 67, as determined in accordance with AMCA 205 by an *approved*, independent testing laboratory and labeled by the manufacturer. The total efficiency of the fan at the design point of operation shall be within 15 percentage points of the maximum total efficiency of the fan.

Exception: The following fans are not required to have a fan efficiency grade:

1. Fans of 5 hp (3.7 kW) or less as follows:
   1.1. Individual fans with a motor nameplate horsepower of 5 hp (3.7 kW) or less, unless Exception 1.2 applies.
   1.2. Multiple fans in series or parallel that have a combined motor nameplate horsepower of 5 hp (3.7 kW) or less and are operated as the functional equivalent of a single fan.
2. Fans that are part of equipment covered in Section C403.3.2.
3. Fans included in an equipment package certified by an *approved agency* for air or energy performance.
4. Powered wall/roof ventilators.
5. Fans outside the scope of AMCA 205.
6. Fans that are intended to operate only during emergency conditions.

C403.8.4 Fractional hp fan motors (Mandatory). Motors for fans that are not less than $^{1}/_{12}$ hp (0.062 kW) and less than 1 hp (0.746 kW) shall be electronically commutated motors or shall have a minimum motor efficiency of 70 percent, rated in accordance with DOE 10 CFR 431. These motors shall have the means to adjust motor speed for either balancing or remote control. The use of belt-driven fans to sheave adjustments for airflow balancing instead of a varying motor speed shall be permitted.

Exceptions: The following motors are not required to comply with this section:

1. Motors in the airstream within fan coils and terminal units that only provide heating to the space served.

2. Motors in space-conditioning equipment that comply with Section C403.3.2 or Sections C403.8.1. through C403.8.3.
3. Motors that comply with Section C405.7.

C403.8.5 Fan control. Controls shall be provided for fans in accordance with Section C403.8.5.1 and as required for specific systems provided in Section C403.

C403.8.5.1 Fan airflow control. Each cooling system listed in Table C403.8.5.1 shall be designed to vary the indoor fan airflow as a function of load and shall comply with the following requirements:

1. Direct expansion (DX) and chilled water cooling units that control the capacity of the mechanical cooling directly based on space temperature shall have not fewer than two stages of fan control. Low or minimum speed shall not be greater than 66 percent of full speed. At low or minimum speed, the fan system shall draw not more than 40 percent of the fan power at full fan speed. Low or minimum speed shall be used during periods of low cooling load and ventilation-only operation.
2. Other units including DX cooling units and chilled water units that control the space temperature by modulating the airflow to the space shall have modulating fan control. Minimum speed shall be not greater than 50 percent of full speed. At minimum speed the fan system shall draw not more than 30 percent of the power at full fan speed. Low or minimum speed shall be used during periods of low cooling load and ventilation-only operation.
3. Units that include an air-side economizer in accordance with Section C403.5 shall have not fewer than two speeds of fan control during economizer operation.

Exceptions:

1. Modulating fan control is not required for chilled water and evaporative cooling units with fan motors of less than 1 hp (0.746 kW) where the units are not used to provide *ventilation air* and the indoor fan cycles with the load.
2. Where the volume of outdoor air required to comply with the ventilation requirements of the *International Mechanical Code* at low speed exceeds the air that would be delivered at the speed defined in Section C403.8.5, the minimum speed shall be selected to provide the required *ventilation air*.

TABLE C403.8.5.1
COOLING SYSTEMS

COOLING SYSTEM TYPE	FAN MOTOR SIZE	MECHANICAL COOLING CAPACITY
DX cooling	Any	≥ 65,000 Btu/h
Chilled water and evaporative cooling	≥ $^1/_4$ hp	Any

For SI: 1 British thermal unit per hour = 0.2931 W; 1 hp = 0.746 kW.

C403.9 Heat rejection equipment. Heat rejection equipment, including air-cooled condensers, dry coolers, open-circuit cooling towers, closed-circuit cooling towers and evaporative condensers, shall comply with this section.

Exception: Heat rejection devices where energy usage is included in the equipment efficiency ratings listed in Tables C403.3.2(6) and C403.3.2(7).

C403.9.1 Fan speed control. Each fan system powered by an individual motor or array of motors with connected power, including the motor service factor, totaling 5 hp (3.7 kW) or more shall have controls and devices configured to automatically modulate the fan speed to control the leaving fluid temperature or condensing temperature and pressure of the heat rejection device. Fan motor power input shall be not more than 30 percent of design wattage at 50 percent of the design airflow.

Exceptions:

1. Fans serving multiple refrigerant or fluid cooling circuits.
2. Condenser fans serving flooded condensers.

C403.9.2 Multiple-cell heat rejection equipment. Multiple-cell heat rejection equipment with variable speed fan drives shall be controlled to operate the maximum number of fans allowed that comply with the manufacturer's requirements for all system components and so that all fans operate at the same fan speed required for the instantaneous cooling duty, as opposed to staged on and off operation. The minimum fan speed shall be the minimum allowable speed of the fan drive system in accordance with the manufacturer's recommendations.

C403.9.3 Limitation on centrifugal fan open-circuit cooling towers. Centrifugal fan open-circuit cooling towers with a combined rated capacity of 1,100 gpm (4164 L/m) or greater at 95°F (35°C) condenser water return, 85°F (29°C) condenser water supply, and 75°F (24°C) outdoor air wet-bulb temperature shall meet the energy efficiency requirement for axial fan open-circuit cooling towers listed in Table C403.3.2(8).

Exception: Centrifugal open-circuit cooling towers that are designed with inlet or discharge ducts or require external sound attenuation.

C403.9.4 Tower flow turndown. Open-circuit cooling towers used on water-cooled chiller systems that are configured with multiple- or variable-speed condenser water pumps shall be designed so that all open- circuit cooling tower cells can be run in parallel with the larger of the flow that is produced by the smallest pump at its minimum expected flow rate or at 50 percent of the design flow for the cell.

C403.9.5 Heat recovery for service water heating. Condenser heat recovery shall be installed for heating or reheating of service hot water provided that the facility operates 24 hours a day, the total installed heat capacity of water-cooled systems exceeds 6,000,000 Btu/hr (1758 kW) of heat rejection, and the design service water heating load exceeds 1,000,000 Btu/h (293 kW).

The required heat recovery system shall have the capacity to provide the smaller of the following:

1. Sixty percent of the peak heat rejection load at design conditions.
2. The preheating required to raise the peak service hot water draw to 85°F (29°C).

Exceptions:

1. Facilities that employ condenser heat recovery for space heating or reheat purposes with a heat recovery design exceeding 30 percent of the peak water-cooled condenser load at design conditions.
2. Facilities that provide 60 percent of their service water heating from site solar or site recovered energy or from other sources.

C403.10 Refrigeration equipment performance. Refrigeration equipment shall have an energy use in kWh/day not greater than the values of Tables C403.10.1(1) and C403.10.1(2) when tested and rated in accordance with AHRI Standard 1200. The energy use shall be verified through certification under an approved certification program or, where a certification program does not exist, the energy use shall be supported by data furnished by the equipment manufacturer.

C403.10.1 Walk-in coolers, walk-in freezers, refrigerated warehouse coolers and refrigerated warehouse freezers (Mandatory). *Refrigerated warehouse* coolers and *refrigerated warehouse freezers* shall comply with this section. *Walk-in coolers* and *walk-in freezers* that are neither site assembled nor site constructed shall comply with the following:

1. Be equipped with automatic door-closers that firmly close walk-in doors that have been closed to within 1 inch (25 mm) of full closure.

   Exception: Automatic closers are not required for doors more than 45 inches (1143 mm) in width or more than 7 feet (2134 mm) in height.

2. Doorways shall have strip doors, curtains, spring-hinged doors or other method of minimizing infiltration when doors are open.
3. *Walk-in coolers* and *refrigerated warehouse coolers* shall contain wall, ceiling, and door insulation of not less than R-25 and *walk-in freezers* and *refrigerated warehouse freezers* shall contain wall, ceiling and door insulation of not less than R-32.

   Exception: Glazed portions of doors or structural members need not be insulated.

4. *Walk-in freezers* shall contain floor insulation of not less than R-28.
5. Transparent reach-in doors for *walk-in freezers* and windows in *walk-in freezer* doors shall be of triple-pane glass, either filled with inert gas or with heat-reflective treated glass.
6. Windows and transparent reach-in doors for *walk-in coolers* shall be of double-pane or triple-pane, inert gas-filled, heat-reflective treated glass.
7. Evaporator fan motors that are less than 1 hp (0.746 kW) and less than 460 volts shall use electronically commutated motors, brushless direct-current motors, or 3-phase motors.
8. Condenser fan motors that are less than 1 hp (0.746 kW) shall use electronically commutated motors, permanent split capacitor-type motors or 3-phase motors.
9. Where antisweat heaters without antisweat heater controls are provided, they shall have a total door rail, glass and frame heater power draw of not more than 7.1 W/ft^2 (76 W/m^2) of door opening for *walk-in freezers* and 3.0 W/ft^2 (32 W/m^2) of door opening for *walk-in coolers*.
10. Where antisweat heater controls are provided, they shall reduce the energy use of the antisweat heater as a function of the relative humidity in the air outside the door or to the condensation on the inner glass pane.
11. Lights in *walk-in coolers*, *walk-in freezers*, *refrigerated warehouse coolers* and *refrigerated warehouse freezers* shall either use light sources with an efficacy of not less than 40 lumens per watt, including ballast losses, or shall use light sources with an efficacy of not less than 40 lumens per watt, including ballast losses, in conjunction with a device that turns off the lights within 15 minutes when the space is not occupied.

C403.10.2 Walk-in coolers and walk-in freezers (Mandatory). Site-assembled or site-constructed *walk-in coolers* and *walk-in freezers* shall comply with the following:

1. Automatic door closers shall be provided that fully close walk-in doors that have been closed to within 1 inch (25 mm) of full closure.

   Exception: Closers are not required for doors more than 45 inches (1143 mm) in width or more than 7 feet (2134 mm) in height.

2. Doorways shall be provided with strip doors, curtains, spring-hinged doors or other method of minimizing infiltration when the doors are open.
3. Walls shall be provided with insulation having a thermal resistance of not less than R-25, ceilings shall be provided with insulation having a thermal resistance of not less than R-25 and doors of *walk-in coolers* and *walk-in freezers* shall be provided with insulation having a thermal resistance of not less than R-32.

   Exception: Insulation is not required for glazed portions of doors or at structural members associated with the walls, ceiling or door frame.

4. The floor of *walk-in freezers* shall be provided with insulation having a thermal resistance of not less than R-28.
5. Transparent reach-in doors for and windows in opaque *walk-in freezer* doors shall be provided with triple-pane glass having the interstitial spaces

**TABLE C403.10.1(1)
MINIMUM EFFICIENCY REQUIREMENTS: COMMERCIAL REFRIGERATION**

EQUIPMENT TYPE	APPLICATION	ENERGY USE LIMITS (kWh per day)[a]	TEST PROCEDURE
Refrigerator with solid doors	Holding Temperature	$0.10 \times V + 2.04$	AHRI 1200
Refrigerator with transparent doors		$0.12 \times V + 3.34$	
Freezers with solid doors		$0.40 \times V + 1.38$	
Freezers with transparent doors		$0.75 \times V + 4.10$	
Refrigerators/freezers with solid doors		the greater of $0.12 \times V + 3.34$ or 0.70	
Commercial refrigerators	Pulldown	$0.126 \times V + 3.51$	

a. V = volume of the chiller or frozen compartment as defined in AHAM-HRF-1.

**TABLE C403.10.1(2)
MINIMUM EFFICIENCY REQUIREMENTS: COMMERCIAL REFRIGERATORS AND FREEZERS**

EQUIPMENT TYPE				ENERGY USE LIMITS (kWh/day)[a, b]	TEST PROCEDURE
Equipment Class[c]	Family Code	Operating Mode	Rating Temperature		
VOP.RC.M	Vertical open	Remote condensing	Medium	$0.82 \times TDA + 4.07$	AHRI 1200
SVO.RC.M	Semivertical open	Remote condensing	Medium	$0.83 \times TDA + 3.18$	
HZO.RC.M	Horizontal open	Remote condensing	Medium	$0.35 \times TDA + 2.88$	
VOP.RC.L	Vertical open	Remote condensing	Low	$2.27 \times TDA + 6.85$	
HZO.RC.L	Horizontal open	Remote condensing	Low	$0.57 \times TDA + 6.88$	
VCT.RC.M	Vertical transparent door	Remote condensing	Medium	$0.22 \times TDA + 1.95$	
VCT.RC.L	Vertical transparent door	Remote condensing	Low	$0.56 \times TDA + 2.61$	
SOC.RC.M	Service over counter	Remote condensing	Medium	$0.51 \times TDA + 0.11$	
VOP.SC.M	Vertical open	Self-contained	Medium	$1.74 \times TDA + 4.71$	
SVO.SC.M	Semivertical open	Self-contained	Medium	$1.73 \times TDA + 4.59$	
HZO.SC.M	Horizontal open	Self-contained	Medium	$0.77 \times TDA + 5.55$	
HZO.SC.L	Horizontal open	Self-contained	Low	$1.92 \times TDA + 7.08$	
VCT.SC.I	Vertical transparent door	Self-contained	Ice cream	$0.67 \times TDA + 3.29$	
VCS.SC.I	Vertical solid door	Self-contained	Ice cream	$0.38 \times V + 0.88$	
HCT.SC.I	Horizontal transparent door	Self-contained	Ice cream	$0.56 \times TDA + 0.43$	
SVO.RC.L	Semivertical open	Remote condensing	Low	$2.27 \times TDA + 6.85$	
VOP.RC.I	Vertical open	Remote condensing	Ice cream	$2.89 \times TDA + 8.7$	
SVO.RC.I	Semivertical open	Remote condensing	Ice cream	$2.89 \times TDA + 8.7$	
HZO.RC.I	Horizontal open	Remote condensing	Ice cream	$0.72 \times TDA + 8.74$	
VCT.RC.I	Vertical transparent door	Remote condensing	Ice cream	$0.66 \times TDA + 3.05$	
HCT.RC.M	Horizontal transparent door	Remote condensing	Medium	$0.16 \times TDA + 0.13$	

(continued)

TABLE C403.10.1(2)—continued
MINIMUM EFFICIENCY REQUIREMENTS: COMMERCIAL REFRIGERATORS AND FREEZERS

EQUIPMENT TYPE				ENERGY USE LIMITS (kWh/day)[a, b]	TEST PROCEDURE
Equipment Class[c]	Family Code	Operating Mode	Rating Temperature		
HCT.RC.L	Horizontal transparent door	Remote condensing	Low	0.34 × TDA + 0.26	AHRI 1200
HCT.RC.I	Horizontal transparent door	Remote condensing	Ice cream	0.4 × TDA + 0.31	
VCS.RC.M	Vertical solid door	Remote condensing	Medium	0.11 × V + 0.26	
VCS.RC.L	Vertical solid door	Remote condensing	Low	0.23 × V + 0.54	
VCS.RC.I	Vertical solid door	Remote condensing	Ice cream	0.27 × V + 0.63	
HCS.RC.M	Horizontal solid door	Remote condensing	Medium	0.11 × V + 0.26	
HCS.RC.L	Horizontal solid door	Remote condensing	Low	0.23 × V + 0.54	
HCS.RC.I	Horizontal solid door	Remote condensing	Ice cream	0.27 × V + 0.63	
HCS.RC.I	Horizontal solid door	Remote condensing	Ice cream	0.27 × V + 0.63	
SOC.RC.L	Service over counter	Remote condensing	Low	1.08 × TDA + 0.22	
SOC.RC.I	Service over counter	Remote condensing	Ice cream	1.26 × TDA + 0.26	
VOP.SC.L	Vertical open	Self-contained	Low	4.37 × TDA + 11.82	
VOP.SC.I	Vertical open	Self-contained	Ice cream	5.55 × TDA + 15.02	
SVO.SC.L	Semivertical open	Self-contained	Low	4.34 × TDA + 11.51	
SVO.SC.I	Semivertical open	Self-contained	Ice cream	5.52 × TDA + 14.63	
HZO.SC.I	Horizontal open	Self-contained	Ice cream	2.44 × TDA + 9.0	
SOC.SC.I	Service over counter	Self-contained	Ice cream	1.76 × TDA + 0.36	
HCS.SC.I	Horizontal solid door	Self-contained	Ice cream	0.38 × V + 0.88	

a. V = Volume of the case, as measured in accordance with Appendix C of AHRI 1200.
b. TDA = Total display area of the case, as measured in accordance with Appendix D of AHRI 1200.
c. Equipment class designations consist of a combination [in sequential order separated by periods (AAA).(BB).(C)] of:

(AAA) An equipment family code where:
VOP = vertical open
SVO = semivertical open
HZO = horizontal open
HCT = horizontal transparent doors
HCS = horizontal solid doors
SOC = service over counter

(BB) An operating mode code:
RC = remote condensing
SC = self-contained

(C) A rating temperature code:
M = medium temperature (38°F)
L = low temperature (0°F)
I = ice-cream temperature (15°F)

For example, "VOP.RC.M" refers to the "vertical-open, remote-condensing, medium-temperature" equipment class.

filled with inert gas or provided with heat-reflective treated glass.

6. Transparent reach-in doors for and windows in opaque *walk-in cooler* doors shall be double-pane heat-reflective treated glass having the interstitial space gas filled.
7. Evaporator fan motors that are less than 1 hp (0.746 kW) and less than 460 volts shall be electronically commutated motors or 3-phase motors.
8. Condenser fan motors that are less than 1 hp (0.746 kW) in capacity shall be of the electroni-

cally commutated or permanent split capacitor-type or shall be 3-phase motors.

Exception: Fan motors in *walk-in coolers* and *walk-in freezers* combined in a single enclosure greater than 3,000 square feet (279 m^2) in floor area are exempt.

9. Antisweat heaters that are not provided with antisweat heater controls shall have a total door rail, glass and frame heater power draw not greater than 7.1 W/ft^2 (76 W/m^2) of door opening for *walk-in freezers*, and not greater than 3.0 W/ft^2 (32 W/m^2) of door opening for *walk-in coolers*.
10. Antisweat heater controls shall be configured to reduce the energy use of the antisweat heater as a function of the relative humidity in the air outside the door or to the condensation on the inner glass pane.
11. Light sources shall have an efficacy of not less than 40 lumens per Watt, including any ballast losses, or shall be provided with a device that automatically turns off the lights within 15 minutes of when the *walk-in cooler* or *walk-in freezer* was last occupied.

C403.10.2.1 Performance standards (Mandatory). Effective January 1, 2020, *walk-in coolers* and *walk-in freezers* shall meet the requirements of Tables C403.10.2.1(1), C403.10.2.1(2) and C403.10.2.1(3).

C403.10.3 Refrigerated display cases (Mandatory). Site-assembled or site-constructed refrigerated display cases shall comply with the following:

1. Lighting and glass doors in refrigerated display cases shall be controlled by one of the following:
   1.1. Time-switch controls to turn off lights during nonbusiness hours. Timed overrides for display cases shall turn the lights on for up to 1 hour and shall automatically time out to turn the lights off.
   1.2. Motion sensor controls on each display case section that reduce lighting power by not less than 50 percent within 3 minutes after the area within the sensor range is vacated.
2. Low-temperature display cases shall incorporate temperature-based defrost termination control with a time-limit default. The defrost cycle shall terminate first on an upper temperature limit breach and second upon a time limit breach.
3. Antisweat heater controls shall reduce the energy use of the antisweat heater as a function of the relative humidity in the air outside the door or to the condensation on the inner glass pane.

C403.10.4 Refrigeration systems. Refrigerated display cases, *walk-in coolers* or *walk-in freezers* that are served by remote compressors and remote condensers not located in a condensing unit, shall comply with Sections C403.10.4.1 and C403.10.4.2.

Exception: Systems where the working fluid in the refrigeration cycle goes through both subcritical and super-critical states (transcritical) or that use ammonia refrigerant are exempt.

TABLE C403.10.2.1(1)
WALK-IN COOLER AND FREEZER DISPLAY DOOR EFFICIENCY REQUIREMENTS[a]

CLASS DESCRIPTOR	CLASS	MAXIMUM ENERGY CONSUMPTION (kWh/day)[a]
Display door, medium temperature	DD, M	$0.04 \times A_{dd} + 0.41$
Display door, low temperature	DD, L	$0.15 \times A_{dd} + 0.29$

a. A_{dd} is the surface area of the display door.

TABLE C403.10.2.1(2)
WALK-IN COOLER AND FREEZER NONDISPLAY DOOR EFFICIENCY REQUIREMENTS[a]

CLASS DESCRIPTOR	CLASS	MAXIMUM ENERGY CONSUMPTION (kWh/day)[a]
Passage door, medium temperature	PD, M	$0.05 \times A_{nd} + 1.7$
Passage door, low temperature	PD, L	$0.14 \times A_{nd} + 4.8$
Freight door, medium temperature	FD, M	$0.04 \times A_{nd} + 1.9$
Freight door, low temperature	FD, L	$0.12 \times A_{nd} + 5.6$

a. A_{nd} is the surface area of the nondisplay door.

TABLE C403.10.2.1(3)
WALK-IN COOLER AND FREEZER REFRIGERATION SYSTEM EFFICIENCY REQUIREMENTS

CLASS DESCRIPTOR	CLASS	MINIMUM ANNUAL WALK-IN ENERGY FACTOR AWEF (Btu/W-h)
Dedicated condensing, medium temperature, indoor system	DC.M.I	5.61
Dedicated condensing, medium temperature, indoor system, > 9,000 Btu/h capacity	DC.M.I, > 9,000	5.61
Dedicated condensing, medium temperature, outdoor system	DC.M.I	7.60
Dedicated condensing, medium temperature, outdoor system, > 9,000 Btu/h capacity	DC.M.I, > 9,000	7.60

C403.10.4.1 Condensers serving refrigeration systems. Fan-powered condensers shall comply with the following:

1. The design *saturated condensing temperatures* for air-cooled condensers shall not exceed the design dry-bulb temperature plus 10°F (5.6°C) for low-*temperature refrigeration systems*, and the design dry- bulb temperature plus 15°F (8°C) for *medium temperature refrigeration systems* where the *saturated condensing temperature* for blend refrigerants shall be determined using the average of liquid and vapor temperatures as converted from the condenser drain pressure.
2. Condenser fan motors that are less than 1 hp (0.75 kW) shall use electronically commutated motors, permanent split-capacitor-type motors or 3-phase motors.
3. Condenser fans for air-cooled condensers, evaporatively cooled condensers, air- or water-cooled fluid coolers or cooling towers shall reduce fan motor demand to not more than 30 percent of design wattage at 50 percent of design air volume, and incorporate one of the following continuous variable speed fan control approaches:
   - 3.1. Refrigeration system condenser control for air-cooled condensers shall use variable setpoint control logic to reset the condensing temperature setpoint in response to ambient dry-bulb temperature.
   - 3.2. Refrigeration system condenser control for evaporatively cooled condensers shall use variable setpoint control logic to reset the condensing temperature setpoint in response to ambient wet-bulb temperature.
4. Multiple fan condensers shall be controlled in unison.
5. The minimum condensing temperature setpoint shall be not greater than 70°F (21°C).

C403.10.4.2 Compressor systems. Refrigeration compressor systems shall comply with the following:

1. Compressors and multiple-compressor system suction groups shall include control systems that use floating suction pressure control logic to reset the target suction pressure temperature based on the temperature requirements of the attached refrigeration display cases or walk-ins.

   Exception: Controls are not required for the following:

   1. Single-compressor systems that do not have variable capacity capability.
   2. Suction groups that have a design saturated suction temperature of 30°F (-1.1°C) or higher, suction groups that comprise the high stage of a two-stage or cascade system, or suction groups that primarily serve chillers for secondary cooling fluids.
2. Liquid subcooling shall be provided for all low-temperature compressor systems with a design cooling capacity equal to or greater than 100,000 Btu/h (29.3 kW) with a design-saturated suction temperature of -10°F (-23°C) or lower. The sub-cooled liquid temperature shall be controlled at a maximum temperature setpoint of 50°F (10°C) at the exit of the subcooler using either compressor economizer (interstage) ports or a separate compressor suction group operating at a saturated suction temperature of 18°F (-7.8°C) or higher.
   - 2.1. Insulation for liquid lines with a fluid operating temperature less than 60°F (15.6°C) shall comply with Table C403.11.3.
3. Compressors that incorporate internal or external crankcase heaters shall provide a means to cycle the heaters off during compressor operation.

C403.11 Construction of HVAC system elements (Mandatory). Ducts, plenums, piping and other elements that are part of an HVAC system shall be constructed and insulated in accordance with Sections C403.11.1 through C403.11.3.1.

C403.11.1 Duct and plenum insulation and sealing (Mandatory). Supply and return air ducts and plenums shall be insulated with not less than R-6 insulation where located in unconditioned spaces and where located outside the building with not less than R-8 insulation in *Climate Zones* 1 through 4 and not less than R-12 insulation in *Climate Zones* 5 through 8. Where located within a building envelope assembly, the duct or plenum shall be separated from the building exterior or unconditioned or exempt spaces by not less than R-8 insulation in *Climate Zones* 1 through 4 and not less than R-12 insulation in *Climate Zones* 5 through 8.

Exceptions:

1. Where located within equipment.
2. Where the design temperature difference between the interior and exterior of the duct or plenum is not greater than 15°F (8°C).

Ducts, air handlers and filter boxes shall be sealed. Joints and seams shall comply with Section 603.9 of the *International Mechanical Code*.

C403.11.2 Duct construction (Mandatory). Ductwork shall be constructed and erected in accordance with the *International Mechanical Code*.

C403.11.2.1 Low-pressure duct systems (Mandatory). Longitudinal and transverse joints, seams and connections of supply and return ducts operating at a static pressure less than or equal to 2 inches water gauge (w.g.) (498 Pa) shall be securely fastened and sealed with welds, gaskets, mastics (adhesives), mastic-plus-embedded-fabric systems or tapes installed in accordance with the manufacturer's instructions. Pressure classifications specific to the duct system shall be clearly indicated on the construction documents in accordance with the *International Mechanical Code*.

Exception: Locking-type longitudinal joints and seams, other than the snap-lock and button-lock types, need not be sealed as specified in this section.

C403.11.2.2 Medium-pressure duct systems (Mandatory). Ducts and plenums designed to operate at a static pressure greater than 2 inches water gauge (w.g.) (498 Pa) but less than 3 inches w.g. (747 Pa) shall be insulated and sealed in accordance with Section C403.11.1. Pressure classifications specific to the duct system shall be clearly indicated on the construction documents in accordance with the *International Mechanical Code*.

C403.11.2.3 High-pressure duct systems (Mandatory). Ducts and plenums designed to operate at static pressures equal to or greater than 3 inches water gauge (747 Pa) shall be insulated and sealed in accordance with Section C403.11.1. In addition, ducts and plenums shall be leak tested in accordance with the SMACNA HVAC Air Duct Leakage Test Manual and shown to have a rate of air leakage (CL) less than or equal to 4.0 as determined in accordance with Equation 4-8.

$$CL = F/P^{0.65} \quad \textbf{(Equation 4-8)}$$

where:

F = The measured leakage rate in cfm per 100 square feet of duct surface.

P = The static pressure of the test.

Documentation shall be furnished by the designer demonstrating that representative sections totaling not less than 25 percent of the duct area have been tested and that all tested sections comply with the requirements of this section.

C403.11.3 Piping insulation (Mandatory). Piping serving as part of a heating or cooling system shall be thermally insulated in accordance with Table C403.11.3.

Exceptions:

1. Factory-installed piping within HVAC equipment tested and rated in accordance with a test procedure referenced by this code.
2. Factory-installed piping within room fan-coils and unit ventilators tested and rated according to AHRI 440 (except that the sampling and variation provisions of Section 6.5 shall not apply) and AHRI 840, respectively.
3. Piping that conveys fluids that have a design operating temperature range between 60°F (15°C) and 105°F (41°C).
4. Piping that conveys fluids that have not been heated or cooled through the use of fossil fuels or electric power.
5. Strainers, control valves, and balancing valves associated with piping 1 inch (25 mm) or less in diameter.
6. Direct buried piping that conveys fluids at or below 60°F (15°C).

C403.11.3.1 Protection of piping insulation (Mandatory). Piping insulation exposed to the weather shall be protected from damage, including that caused by sun-

TABLE C403.11.3
MINIMUM PIPE INSULATION THICKNESS (in inches)[a, c]

FLUID OPERATING TEMPERATURE RANGE AND USAGE (°F)	INSULATION CONDUCTIVITY		NOMINAL PIPE OR TUBE SIZE (inches)				
	Conductivity Btu • in./(h • ft² • °F)[b]	Mean Rating Temperature, °F	< 1	1 to < $1^1/_2$	$1^1/_2$ to < 4	4 to < 8	≥ 8
> 350	0.32 – 0.34	250	4.5	5.0	5.0	5.0	5.0
251 – 350	0.29 – 0.32	200	3.0	4.0	4.5	4.5	4.5
201 – 250	0.27 – 0.30	150	2.5	2.5	2.5	3.0	3.0
141 – 200	0.25 – 0.29	125	1.5	1.5	2.0	2.0	2.0
105 – 140	0.21 – 0.28	100	1.0	1.0	1.5	1.5	1.5
40 – 60	0.21 – 0.27	75	0.5	0.5	1.0	1.0	1.0
< 40	0.20 – 0.26	50	0.5	1.0	1.0	1.0	1.5

For SI: 1 inch = 25.4 mm, °C = [(°F) - 32]/1.8.

a. For piping smaller than $1^1/_2$ inches and located in partitions within conditioned spaces, reduction of these thicknesses by 1 inch shall be permitted (before thickness adjustment required in footnote b) but not to a thickness less than 1 inch.

b. For insulation outside the stated conductivity range, the minimum thickness (T) shall be determined as follows:

$T = r[(1 + t/r)^{K/k} - 1]$

where:

T = minimum insulation thickness,

r = actual outside radius of pipe,

t = insulation thickness listed in the table for applicable fluid temperature and pipe size,

K = conductivity of alternate material at mean rating temperature indicated for the applicable fluid temperature (Btu • in/h • ft² • °F) and

k = the upper value of the conductivity range listed in the table for the applicable fluid temperature.

c. For direct-buried heating and hot water system piping, reduction of these thicknesses by $1^1/_2$ inches (38 mm) shall be permitted (before thickness adjustment required in footnote b but not to thicknesses less than 1 inch.

light, moisture, equipment maintenance and wind, and shall provide shielding from solar radiation that can cause degradation of the material. Adhesive tape shall not be permitted.

C403.12 Mechanical systems located outside of the building thermal envelope (Mandatory). Mechanical systems providing heat outside of the thermal envelope of a building shall comply with Sections C403.12.1 through C403.12.3.

C403.12.1 Heating outside a building. Systems installed to provide heat outside a building shall be radiant systems.

Such heating systems shall be controlled by an occupancy sensing device or a timer switch, so that the system is automatically de-energized when occupants are not present.

C403.12.2 Snow- and ice-melt system controls. Snow- and ice-melting systems shall include automatic controls configured to shut off the system when the pavement temperature is above 50°F (10°C) and precipitation is not falling, and an automatic or manual control that is configured to shut off when the outdoor temperature is above 40°F (4°C).

C403.12.3 Freeze protection system controls. Freeze protection systems, such as heat tracing of outdoor piping and heat exchangers, including self-regulating heat tracing, shall include automatic controls configured to shut off the systems when outdoor air temperatures are above 40°F (4°C) or when the conditions of the protected fluid will prevent freezing.

SECTION C404 SERVICE WATER HEATING (MANDATORY)

C404.1 General. This section covers the minimum efficiency of, and controls for, service water-heating equipment and insulation of service hot water piping.

C404.2 Service water-heating equipment performance efficiency. Water-heating equipment and hot water storage tanks shall meet the requirements of Table C404.2. The efficiency shall be verified through data furnished by the manufacturer of the equipment or through certification under an *approved* certification program. Water-heating equipment intended to be used to provide space heating shall meet the applicable provisions of Table C404.2.

C404.2.1 High input service water-heating systems. Gas-fired water-heating equipment installed in new buildings shall be in compliance with this section. Where a singular piece of water-heating equipment serves the entire building and the input rating of the equipment is 1,000,000 Btu/h (293 kW) or greater, such equipment shall have a thermal efficiency, E_t, of not less than 90 percent. Where multiple pieces of water-heating equipment serve the building and the combined input rating of the water-heating equipment is 1,000,000 Btu/h (293 kW) or greater, the combined input-capacity-weighted-average thermal efficiency, E_t, shall be not less than 90 percent.

Exceptions:

1. Where not less than 25 percent of the annual *service water-heating* requirement is provided by *on-site renewable energy* or site-recovered energy, the minimum thermal efficiency requirements of this section shall not apply.
2. The input rating of water heaters installed in individual dwelling units shall not be required to be included in the total input rating of *service water-heating* equipment for a building.
3. The input rating of water heaters with an input rating of not greater than 100,000 Btu/h (29.3 kW) shall not be required to be included in the total input rating of *service water-heating* equipment for a building.

C404.3 Heat traps for hot water storage tanks. Storage tank-type water heaters and hot water storage tanks that have vertical water pipes connecting to the inlet and outlet of the tank shall be provided with integral heat traps at those inlets and outlets or shall have pipe-configured heat traps in the piping connected to those inlets and outlets. Tank inlets and outlets associated with solar water heating system circulation loops shall not be required to have heat traps.

C404.4 Insulation of piping. Piping from a water heater to the termination of the heated water fixture supply pipe shall be insulated in accordance with Table C403.11.3. On both the inlet and outlet piping of a storage water heater or heated water storage tank, the piping to a heat trap or the first 8 feet (2438 mm) of piping, whichever is less, shall be insulated. Piping that is heat traced shall be insulated in accordance with Table C403.11.3 or the heat trace manufacturer's instructions. Tubular pipe insulation shall be installed in accordance with the insulation manufacturer's instructions. Pipe insulation shall be continuous except where the piping passes through a framing member. The minimum insulation thickness requirements of this section shall not supersede any greater insulation thickness requirements necessary for the protection of piping from freezing temperatures or the protection of personnel against external surface temperatures on the insulation.

Exception: Tubular pipe insulation shall not be required on the following:

1. The tubing from the connection at the termination of the fixture supply piping to a plumbing fixture or plumbing appliance.
2. Valves, pumps, strainers and threaded unions in piping that is 1 inch (25 mm) or less in nominal diameter.
3. Piping from user-controlled shower and bath mixing valves to the water outlets.
4. Cold-water piping of a demand recirculation water system.
5. Tubing from a hot drinking-water heating unit to the water outlet.
6. Piping at locations where a vertical support of the piping is installed.
7. Piping surrounded by building insulation with a thermal resistance (*R*-value) of not less than R-3.

C404.5 Heated water supply piping. Heated water supply piping shall be in accordance with Section C404.5.1 or C404.5.2. The flow rate through $^1/_4$-inch (6.4 mm) piping shall be not greater than 0.5 gpm (1.9 L/m). The flow rate

TABLE C404.2
MINIMUM PERFORMANCE OF WATER-HEATING EQUIPMENT

EQUIPMENT TYPE	SIZE CATEGORY (input)	SUBCATEGORY OR RATING CONDITION	PERFORMANCE REQUIRED[a, b]	TEST PROCEDURE
Water heaters, electric	≤ 12 kW[d]	Tabletop[e], ≥ 20 gallons and ≤ 120 gallons	$0.93 - 0.00132V$, EF	DOE 10 CFR Part 430
		Resistance ≥ 20 gallons and ≤ 55 gallons	$0.960 - 0.0003V$, EF	
		Grid-enabled[f] > 75 gallons and ≤ 120 gallons	$1.061 - 0.00168V$, EF	
	> 12 kW	Resistance	$(0.3 + 27/V_m)$, %/h	ANSI Z21.10.3
	≤ 24 amps and ≤ 250 volts	Heat pump > 55 gallons and ≤ 120 gallons	$2.057 - 0.00113V$, EF	DOE 10 CFR Part 430
Storage water heaters, gas	≤ 75,000 Btu/h	≥ 20 gallons and ≤ 55 gallons	$0.675 - 0.0015V$, EF	DOE 10 CFR Part 430
		> 55 gallons and ≤ 100 gallons	$0.8012 - 0.00078V$, EF	
	> 75,000 Btu/h and ≤ 155,000 Btu/h	< 4,000 Btu/h/gal	80% E_t $(Q/800 + 110\sqrt{V})$SL, Btu/h	ANSI Z21.10.3
	> 155,000 Btu/h	< 4,000 Btu/h/gal	80% E_t $(Q/800 + 110\sqrt{V})$SL, Btu/h	
Instantaneous water heaters, gas	> 50,000 Btu/h and < 200,000 Btu/h[c]	≥ 4,000 Btu/h/gal and < 2 gal	$0.82 - 0.0019V$, EF	DOE 10 CFR Part 430
	≥ 200,000 Btu/h	≥ 4,000 Btu/h/gal and < 10 gal	80% E_t	ANSI Z21.10.3
	≥ 200,000 Btu/h	≥ 4,000 Btu/h/gal and ≥ 10 gal	80% E_t $(Q/800 + 110\sqrt{V})$SL, Btu/h	
Storage water heaters, oil	≤ 105,000 Btu/h	≥ 20 gal and ≤ 50 gallons	$0.68 - 0.0019V$, EF	DOE 10 CFR Part 430
	≥ 105,000 Btu/h	< 4,000 Btu/h/gal	80% E_t $(Q/800 + 110\sqrt{V})$SL, Btu/h	ANSI Z21.10.3
Instantaneous water heaters, oil	≤ 210,000 Btu/h	≥ 4,000 Btu/h/gal and < 2 gal	$0.59 - 0.0019V$, EF	DOE 10 CFR Part 430
	> 210,000 Btu/h	≥ 4,000 Btu/h/gal and < 10 gal	80% E_t	ANSI Z21.10.3
	> 210,000 Btu/h	≥ 4,000 Btu/h/gal and ≥ 10 gal	78% E_t $(Q/800 + 110\sqrt{V})$SL, Btu/h	
Hot water supply boilers, gas and oil	≥ 300,000 Btu/h and < 12,500,000 Btu/h	≥ 4,000 Btu/h/gal and < 10 gal	80% E_t	ANSI Z21.10.3
Hot water supply boilers, gas	≥ 300,000 Btu/h and < 12,500,000 Btu/h	≥ 4,000 Btu/h/gal and ≥ 10 gal	80% E_t $(Q/800 + 110\sqrt{V})$SL, Btu/h	
Hot water supply boilers, oil	> 300,000 Btu/h and < 12,500,000 Btu/h	> 4,000 Btu/h/gal and > 10 gal	78% E_t $(Q/800 + 110\sqrt{V})$SL, Btu/h	
Pool heaters, gas and oil	All	—	82% E_t	ASHRAE 146

(continued)

TABLE C404.2—continued
MINIMUM PERFORMANCE OF WATER-HEATING EQUIPMENT

EQUIPMENT TYPE	SIZE CATEGORY (input)	SUBCATEGORY OR RATING CONDITION	PERFORMANCE REQUIRED[a, b]	TEST PROCEDURE
Heat pump pool heaters	All	—	4.0 COP	AHRI 1160
Unfired storage tanks	All	—	Minimum insulation requirement R-12.5 (h • ft^2 • °F)/Btu	(none)

For SI: 1 foot = 304.8 mm, 1 square foot = 0.0929 m^2,°C = [(°F) - 32]/1.8, 1 British thermal unit per hour = 0.2931 W, 1 gallon = 3.785 L, 1 British thermal unit per hour per gallon = 0.078 W/L.

a. Energy factor (EF) and thermal efficiency (E_t) are minimum requirements. In the EF equation, *V* is the rated volume in gallons.

b. Standby loss (SL) is the maximum Btu/h based on a nominal 70°F temperature difference between stored water and ambient requirements. In the SL equation, *Q* is the nameplate input rate in Btu/h. In the equations for electric water heaters, V is the rated volume in gallons and V_m is the measured volume in gallons. In the SL equation for oil and gas water heaters and boilers, *V* is the rated volume in gallons.

c. Instantaneous water heaters with input rates below 200,000 Btu/h shall comply with these requirements where the water heater is designed to heat water to temperatures 180°F or higher.

d. Electric water heaters with an input rating of 12 kW (40,950 Btu/h) or less that are designed to heat water to temperatures of 180°F or greater shall comply with the requirements for electric water heaters that have an input rating greater than 12 kW (40,950 Btu/h).

e A tabletop water heater is a water heater that is enclosed in a rectangular cabinet with a flat top surface not more than 3 feet in height.

f. A grid-enabled water heater is an electric resistance water heater that meets all of the following:
1. Has a rated storage tank volume of more than 75 gallons.
2. Was manufactured on or after April 16, 2015.
3. Is equipped at the point of manufacture with an activation lock.
4. Bears a permanent label applied by the manufacturer that complies with all of the following:
   4.1. Is made of material not adversely affected by water.
   4.2. Is attached by means of nonwater-soluble adhesive.
   4.3. Advises purchasers and end users of the intended and appropriate use of the product with the following notice printed in 16.5 point Arial Narrow Bold font: "IMPORTANT INFORMATION: This water heater is intended only for use as part of an electric thermal storage or demand response program. It will not provide adequate hot water unless enrolled in such a program and activated by your utility company or another program operator. Confirm the availability of a program in your local area before purchasing or installing this product."

through $^5/_{16}$-inch (7.9 mm) piping shall be not greater than 1 gpm (3.8 L/m). The flow rate through $^3/_8$-inch (9.5 mm) piping shall be not greater than 1.5 gpm (5.7 L/m).

C404.5.1 Maximum allowable pipe length method. The maximum allowable piping length from the nearest source of heated water to the termination of the fixture supply pipe shall be in accordance with the following. Where the piping contains more than one size of pipe, the largest size of pipe within the piping shall be used for determining the maximum allowable length of the piping in Table C404.5.1.

1. For a public lavatory faucet, use the "Public lavatory faucets" column in Table C404.5.1.
2. For all other plumbing fixtures and plumbing appliances, use the "Other fixtures and appliances" column in Table C404.5.1.

C404.5.2 Maximum allowable pipe volume method. The water volume in the piping shall be calculated in accordance with Section C404.5.2.1. Water heaters, circulating water systems and heat trace temperature maintenance systems shall be considered to be sources of heated water.

The volume from the nearest source of heated water to the termination of the fixture supply pipe shall be as follows:

1. For a public lavatory faucet: not more than 2 ounces (0.06 L).
2. For other plumbing fixtures or plumbing appliances; not more than 0.5 gallon (1.89 L).

C404.5.2.1 Water volume determination. The volume shall be the sum of the internal volumes of pipe, fittings, valves, meters and manifolds between the nearest source of heated water and the termination of the fixture supply pipe. The volume in the piping shall be determined from the "Volume" column in Table C404.5.1. The volume contained within fixture shutoff valves, within flexible water supply connectors to a fixture fitting and within a fixture fitting shall not be included in the water volume determination. Where heated water is supplied by a recirculating system or heat-traced piping, the volume shall include the portion of the fitting on the branch pipe that supplies water to the fixture.

C404.6 Heated-water circulating and temperature maintenance systems. Heated-water circulation systems shall be in accordance with Section C404.6.1. Heat trace temperature maintenance systems shall be in accordance with Section C404.6.2. Controls for hot water storage shall be in accordance with Section C404.6.3. Automatic controls, temperature sensors and pumps shall be in a location with *access*. Manual controls shall be in a location with *ready access*.

C404.6.1 Circulation systems. Heated-water circulation systems shall be provided with a circulation pump. The system return pipe shall be a dedicated return pipe or a cold water supply pipe. Gravity and thermo-syphon circulation systems shall be prohibited. Controls for circulating hot water system pumps shall start the pump based on the identification of a demand for hot water within the occupancy. The controls shall automatically turn off the pump

TABLE C404.5.1
PIPING VOLUME AND MAXIMUM PIPING LENGTHS

NOMINAL PIPE SIZE (inches)	VOLUME (liquid ounces per foot length)	MAXIMUM PIPING LENGTH (feet)	
		Public lavatory faucets	Other fixtures and appliances
$^1/_4$	0.33	6	50
$^5/_{16}$	0.5	4	50
$^3/_8$	0.75	3	50
$^1/_2$	1.5	2	43
$^5/_8$	2	1	32
$^3/_4$	3	0.5	21
$^7/_8$	4	0.5	16
1	5	0.5	13
$1^1/_4$	8	0.5	8
$1^1/_2$	11	0.5	6
2 or larger	18	0.5	4

For SI: 1 inch = 25.4 mm, 1 foot = 304.8 mm, 1 liquid ounce = 0.030 L, 1 gallon = 128 ounces.

when the water in the circulation loop is at the desired temperature and when there is not a demand for hot water.

C404.6.2 Heat trace systems. Electric heat trace systems shall comply with IEEE 515.1. Controls for such systems shall be able to automatically adjust the energy input to the heat tracing to maintain the desired water temperature in the piping in accordance with the times when heated water is used in the occupancy. Heat trace shall be arranged to be turned off automatically when there is not a demand for hot water.

C404.6.3 Controls for hot water storage. The controls on pumps that circulate water between a water heater and a heated-water storage tank shall limit operation of the pump from heating cycle startup to not greater than 5 minutes after the end of the cycle.

C404.7 Demand recirculation controls. Demand recirculation water systems shall have controls that comply with both of the following:

1. The controls shall start the pump upon receiving a signal from the action of a user of a fixture or appliance, sensing the presence of a user of a fixture or sensing the flow of hot or tempered water to a fixture fitting or appliance.
2. The controls shall limit the temperature of the water entering the cold-water piping to not greater than 104°F (40°C).

C404.8 Drain water heat recovery units. Drain water heat recovery units shall comply with CSA B55.2. Potable water-side pressure loss shall be less than 10 psi (69 kPa) at maximum design flow. For *Group R* occupancies, the efficiency of drain water heat recovery unit efficiency shall be in accordance with CSA B55.1.

C404.9 Energy consumption of pools and permanent spas (Mandatory). The energy consumption of pools and permanent spas shall be controlled by the requirements in Sections C404.9.1 through C404.9.3.

C404.9.1 Heaters. The electric power to all heaters shall be controlled by an on-off switch that is an integral part of the heater, mounted on the exterior of the heater, or external to and within 3 feet (914 mm) of the heater in a location with *ready access*. Operation of such switch shall not change the setting of the heater thermostat. Such switches shall be in addition to a circuit breaker for the power to the heater. Gas-fired heaters shall not be equipped with continuously burning ignition pilots.

C404.9.2 Time switches. Time switches or other control methods that can automatically turn off and on heaters and pump motors according to a preset schedule shall be installed for heaters and pump motors. Heaters and pump motors that have built-in time switches shall be in compliance with this section.

Exceptions:

1. Where public health standards require 24-hour pump operation.
2. Pumps that operate solar- and waste-heat-recovery pool heating systems.

C404.9.3 Covers. Outdoor heated pools and outdoor permanent spas shall be provided with a vapor-retardant cover or other *approved* vapor-retardant means.

Exception: Where more than 75 percent of the energy for heating, computed over an operating season of not fewer than 3 calendar months, is from site-recovered energy such as from a heat pump or on-site renewable energy system, covers or other vapor-retardant means shall not be required.

C404.10 Energy consumption of portable spas (Mandatory). The energy consumption of electric-powered portable spas shall be controlled by the requirements of APSP 14.

SECTION C405
ELECTRICAL POWER AND LIGHTING SYSTEMS

C405.1 General (Mandatory). This section covers lighting system controls, the maximum lighting power for interior and exterior applications and electrical energy consumption.

Dwelling units within multifamily buildings shall comply with Section R404.1. All other *dwelling units* shall comply with Section R404.1, or with Sections C405.2.4 and C405.3. *Sleeping units* shall comply with Section C405.2.4, and with Section R404.1 or C405.3. Lighting installed in walk-in coolers, walk-in freezers, refrigerated warehouse coolers and refrigerated warehouse freezers shall comply with the lighting requirements of Section C403.10.1 or C403.10.2.

C405.2 Lighting controls (Mandatory). Lighting systems shall be provided with controls that comply with one of the following.

1. Lighting controls as specified in Sections C405.2.1 through C405.2.6.
2. Luminaire level lighting controls (LLLC) and lighting controls as specified in Sections C405.2.1, C405.2.4 and C405.2.5. The LLLC luminaire shall be independently capable of:
   - 2.1. Monitoring occupant activity to brighten or dim lighting when occupied or unoccupied, respectively.
   - 2.2. Monitoring ambient light, both electric light and daylight, and brighten or dim artificial light to maintain desired light level.
   - 2.3. For each control strategy, configuration and reconfiguration of performance parameters including; bright and dim setpoints, timeouts, dimming fade rates, sensor sensitivity adjustments, and wireless zoning configurations.

Exceptions: Lighting controls are not required for the following:

1. Areas designated as security or emergency areas that are required to be continuously lighted.
2. Interior exit stairways, interior exit ramps and exit passageways.
3. Emergency egress lighting that is normally off.

C405.2.1 Occupant sensor controls. Occupant *sensor controls* shall be installed to control lights in the following space types:

1. Classrooms/lecture/training rooms.
2. Conference/meeting/multipurpose rooms.
3. Copy/print rooms.
4. Lounges/breakrooms.
5. Enclosed offices.
6. Open plan office areas.
7. Restrooms.
8. Storage rooms.
9. Locker rooms.
10. Other spaces 300 square feet (28 m^2) or less that are enclosed by floor-to-ceiling height partitions.
11. Warehouse storage areas.

C405.2.1.1 Occupant sensor control function. Occupant sensor controls in warehouses shall comply with Section C405.2.1.2. Occupant sensor controls in open plan office areas shall comply with Section C405.2.1.3. Occupant sensor controls for all other spaces specified in Section C405.2.1 shall comply with the following:

1. They shall automatically turn off lights within 20 minutes after all occupants have left the space.
2. They shall be manual on or controlled to automatically turn on the lighting to not more than 50-percent power.

   Exception: Full automatic-on controls shall be permitted to control lighting in public corridors, stairways, restrooms, primary building entrance areas and lobbies, and areas where manual-on operation would endanger the safety or security of the room or building occupants.

3. They shall incorporate a manual control to allow occupants to turn off lights.

C405.2.1.2 Occupant sensor control function in warehouses. In warehouses, the lighting in aisleways and open areas shall be controlled with occupant sensors that automatically reduce lighting power by not less than 50 percent when the areas are unoccupied. The occupant sensors shall control lighting in each aisleway independently and shall not control lighting beyond the aisleway being controlled by the sensor.

C405.2.1.3 Occupant sensor control function in open plan office areas. Occupant sensor controls in open plan office spaces less than 300 square feet (28 m^2) in area shall comply with Section C405.2.1.1. Occupant sensor controls in all other open plan office spaces shall comply with all of the following:

1. The controls shall be configured so that general lighting can be controlled separately in control zones with floor areas not greater than 600 square feet (55 m^2) within the open plan office space.
2. The controls shall automatically turn off general lighting in all control zones within 20 minutes after all occupants have left the open plan office space.
3. The controls shall be configured so that general lighting power in each control zone is reduced by not less than 80 percent of the full zone general lighting power in a reasonably uniform illumination pattern within 20 minutes of all occupants leaving that control zone. Control functions that switch control zone lights completely off when the zone is vacant meet this requirement.
4. The controls shall be configured such that any daylight responsive control will activate open plan office space general lighting or control zone general lighting only when occupancy for the same area is detected.

C405.2.2 Time-switch controls. Each area of the building that is not provided with *occupant sensor controls* complying with Section C405.2.1.1 shall be provided with *time-switch controls* complying with Section C405.2.2.1.

Exception: Where a *manual control* provides light reduction in accordance with Section C405.2.2.2, *time-switch controls* shall not be required for the following:

1. Spaces where patient care is directly provided.
2. Spaces where an automatic shutoff would endanger occupant safety or security.
3. Lighting intended for continuous operation.
4. Shop and laboratory classrooms.

C405.2.2.1 Time-switch control function. Each space provided with *time-switch controls* shall be provided with a *manual control* for light reduction in accordance with Section C405.2.2.2. Time-switch *controls* shall include an override switching device that complies with the following:

1. Have a minimum 7-day clock.
2. Be capable of being set for seven different day types per week.
3. Incorporate an automatic holiday "shutoff" feature, which turns off all controlled lighting loads for not fewer than 24 hours and then resumes normally scheduled operations.
4. Have program backup capabilities, which prevent the loss of program and time settings for not fewer than 10 hours, if power is interrupted.
5. Include an override switch that complies with the following:
    5.1. The override switch shall be a manual control.
    5.2. The override switch, when initiated, shall permit the controlled lighting to remain on for not more than 2 hours.
    5.3. Any individual override switch shall control the lighting for an area not larger than 5,000 square feet (465 m^2).

Exceptions:

1. Within mall concourses, auditoriums, sales areas, manufacturing facilities and sports arenas:
    1.1. The time limit shall be permitted to be greater than 2 hours, provided that the switch is a captive key device.
    1.2. The area controlled by the override switch shall not be limited to 5,000 square feet (465 m^2) provided that such area is less than 20,000 square feet (1860 m^2).
2. Where provided with *manual control*, the following areas are not required to have light reduction control:
    2.1. Spaces that have only one luminaire with a rated power of less than 100 watts.
    2.2. Spaces that use less than 0.6 watts per square foot (6.5 W/m^2).
    2.3. Corridors, lobbies, electrical rooms and or mechanical rooms.

C405.2.2.2 Light-reduction controls. Spaces required to have light-reduction controls shall have a *manual control* that allows the occupant to reduce the connected lighting load in a reasonably uniform illumination pattern by not less than 50 percent. Lighting reduction shall be achieved by one of the following or another *approved* method:

1. Controlling all lamps or luminaires.
2. Dual switching of alternate rows of luminaires, alternate luminaires or alternate lamps.
3. Switching the middle lamp luminaires independently of the outer lamps.
4. Switching each luminaire or each lamp.

Exception: Light reduction controls are not required in *daylight zones* with *daylight responsive controls* complying with Section C405.2.3.

C405.2.3 Daylight-responsive controls. *Daylight-responsive controls* complying with Section C405.2.3.1 shall be provided to control the electric lights within *daylight zones* in the following spaces:

1. Spaces with a total of more than 150 watts of *general lighting* within sidelit zones complying with Section C405.2.3.2 *General lighting* does not include lighting that is required to have specific application control in accordance with Section C405.2.4.
2. Spaces with a total of more than 150 watts of *general lighting* within toplit zones complying with Section C405.2.3.3.

Exceptions: Daylight responsive controls are not required for the following:

1. Spaces in health care facilities where patient care is directly provided.
2. Lighting that is required to have specific application control in accordance with Section C405.2.4.
3. Sidelit zones on the first floor above grade in Group A-2 and Group M occupancies.
4. New buildings where the total connected lighting power calculated in accordance with Section C405.3.1 is not greater than the adjusted interior lighting power allowance (LPA_{adj}) calculated in accordance with Equation 4-9:

$$LPA_{adj} = [LPA_{norm} \times (1.0 - 0.4 \times UDZFA / TBFA)] \quad \textbf{(Equation 4-9)}$$

where:

LPA_{adj} = Adjusted building interior lighting power allowance in watts.

LPA_{norm} = Normal building lighting power allowance in watts calculated in accordance with Section C405.3.2 and reduced in accordance with Section C406.3 where Option 2 of Section C406.1 is used to comply with the requirements of Section C406.

UDZFA = Uncontrolled daylight zone floor area is the sum of all sidelit and toplit zones, calculated in accordance with Sections C405.2.3.2 and C405.2.3.3, that do not have daylight responsive controls.

TBFA = Total building floor area is the sum of all floor areas included in the lighting power allowance calculation in Section C405.3.2.

C405.2.3.1 Daylight-responsive control function. Where required, *daylight-responsive controls* shall be provided within each space for control of lights in that space and shall comply with all of the following:

1. Lights in *toplit* zones in accordance with Section C405.2.3.3 shall be controlled independently of lights in sidelit zones in accordance with Section C405.2.3.2.
2. *Daylight responsive controls* within each space shall be configured so that they can be calibrated from within that space by authorized personnel.
3. Calibration mechanisms shall be in a location with *ready access.*
4. Where located in offices, classrooms, laboratories and library reading rooms, *daylight responsive controls* shall dim lights continuously from full light output to 15 percent of full light output or lower.
5. *Daylight responsive controls* shall be configured to completely shut off all controlled lights.
6. Lights in *sidelit zones* in accordance with Section C405.2.3.2 facing different cardinal orientations [within 45 degrees (0.79 rad) of due north, east, south, west] shall be controlled independently of each other.

Exception: Up to 150 watts of lighting in each space is permitted to be controlled together with lighting in a daylight zone facing a different cardinal orientation.

C405.2.3.2 Sidelit zone. The sidelit zone is the floor area adjacent to vertical *fenestration* that complies with all of the following:

1. Where the fenestration is located in a wall, the sidelit zone shall extend laterally to the nearest full-height wall, or up to 1.0 times the height from the floor to the top of the fenestration, and longitudinally from the edge of the fenestration to the nearest full-height wall, or up to 2 feet (610 mm), whichever is less, as indicated in Figure C405.2.3.2.
2. The area of the fenestration is not less than 24 square feet (2.23 m^2).
3. The distance from the fenestration to any building or geological formation that would block *access* to daylight is greater than the height from the bottom of the fenestration to the top of the building or geologic formation.
4. The visible transmittance of the fenestration is not less than 0.20.

C405.2.3.3 Toplit zone. The *toplit* zone is the floor area underneath a roof fenestration assembly that complies with all of the following:

1. The *toplit* zone shall extend laterally and longitudinally beyond the edge of the roof fenestration assembly to the nearest obstruction that is taller than 0.7 times the ceiling height, or up to 0.7 times the ceiling height, whichever is less, as indicated in Figure C405.2.3.3(1).
2. Where the fenestration is located in a rooftop monitor, the toplit zone shall extend laterally to the nearest obstruction that is taller than 0.7 times the ceiling height, or up to 1.0 times the height from the floor to the bottom of the fenestration, whichever is less, and longitudinally from the edge of the fenestration to the nearest obstruction that is taller than 0.7 times the ceiling height, or up to 0.25 times the height from the floor to the bottom of the fenestration, whichever is less, as indicated in Figures C405.2.3.3(2) and C405.2.3.3(3).
3. Direct sunlight is not blocked from hitting the roof fenestration assembly at the peak solar angle on the summer solstice by buildings or geological formations.
4. The product of the visible transmittance of the roof fenestration assembly and the area of the rough opening of the roof fenestration assembly divided by the area of the *toplit* zone is not less than 0.008.

C405.2.4 Specific application controls. Specific application controls shall be provided for the following:

1. The following lighting shall be controlled by an occupant sensor complying with Section C405.2.1.1 or a time-switch control complying with Section C405.2.2.1. In addition, a manual control shall be provided to control such lighting separately from the general lighting in the space:
   - 1.1. Display and accent.
   - 1.2. Lighting in display cases.
   - 1.3. Supplemental task lighting, including permanently installed under-shelf or under-cabinet lighting.
   - 1.4. Lighting equipment that is for sale or demonstration in lighting education.
2. *Sleeping units* shall have control devices or systems that are configured to automatically switch off all

permanently installed luminaires and switched receptacles within 20 minutes after all occupants have left the unit.

Exceptions:

1. Lighting and switched receptacles controlled by card key controls.
2. Spaces where patient care is directly provided.

3. Permanently installed luminaires within *dwelling units* shall be provided with controls complying with Section C405.2.1.1 or C405.2.2.2.
4. Lighting for nonvisual applications, such as plant growth and food warming, shall be controlled by a time switch control complying with Section C405.2.2.1 that is independent of the controls for other lighting within the room or space.

C405.2.5 Manual controls. Where required by this code, manual controls for lights shall comply with the following:

1. They shall be in a location with *ready access* to occupants.
2. They shall be located where the controlled lights are visible, or shall identify the area served by the lights and indicate their status.

C405.2.6 Exterior lighting controls. Exterior lighting systems shall be provided with controls that comply with Sections C405.2.6.1 through C405.2.6.4. Decorative lighting systems shall comply with Sections C405.2.6.1, C405.2.6.2 and C405.2.6.4.

Exceptions:

1. Lighting for covered vehicle entrances and exits from buildings and parking structures where required for eye adaptation.
2. Lighting controlled from within dwelling units.

C405.2.6.1 Daylight shutoff. Lights shall be automatically turned off when daylight is present and satisfies the lighting needs.

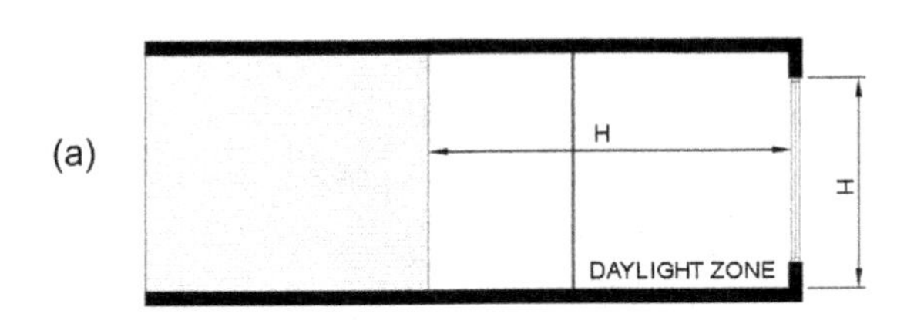

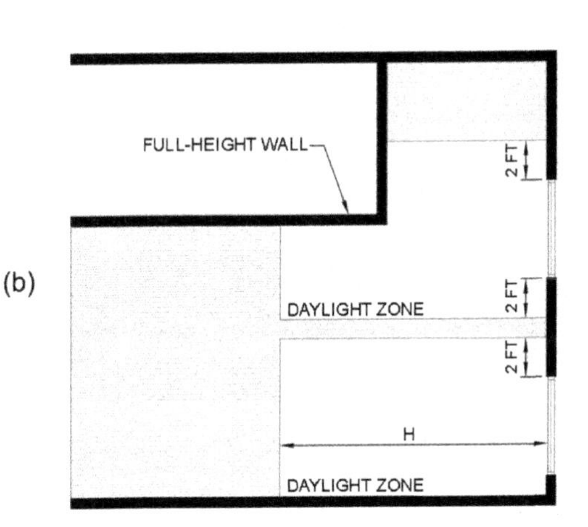

(a) Section view
(b) Plan view of daylight zone under a rooftop monitor

FIGURE C405.2.3.2
SIDELIT ZONE

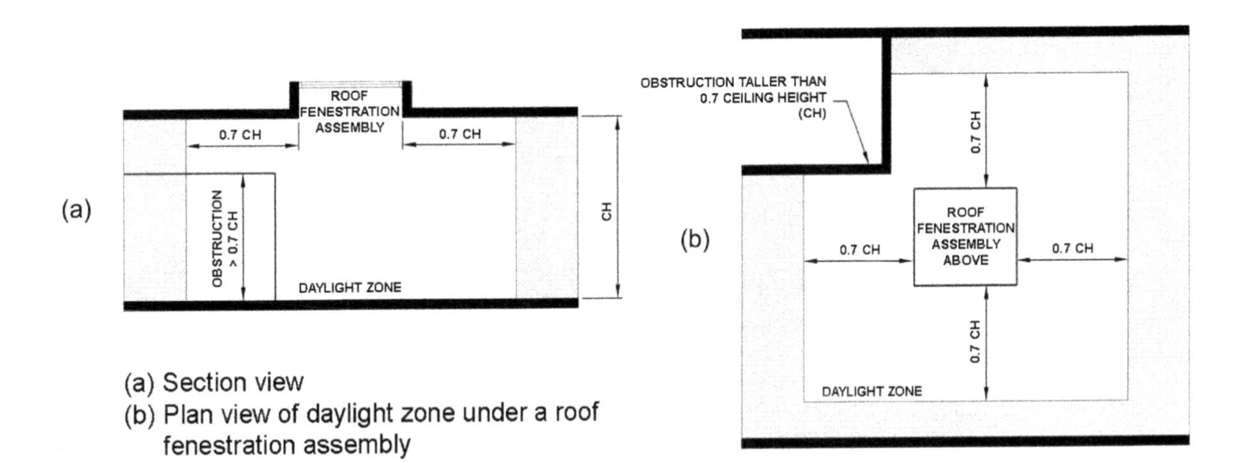

(a) Section view
(b) Plan view of daylight zone under a roof fenestration assembly

FIGURE C405.2.3.3(1)
TOPLIT ZONE

C405.2.6.2 Decorative lighting shutoff. Building facade and landscape lighting shall automatically shut off from not later than 1 hour after business closing to not earlier than 1 hour before business opening.

C405.2.6.3 Lighting setback. Lighting that is not controlled in accordance with Section C405.2.6.2 shall be controlled so that the total wattage of such lighting is automatically reduced by not less than 30 percent by selectively switching off or dimming luminaires at one of the following times:

1. From not later than midnight to not earlier than 6 a.m.
2. From not later than one hour after business closing to not earlier than one hour before business opening.
3. During any time where activity has not been detected for 15 minutes or more.

C405.2.6.4 Exterior time-switch control function. Time-switch controls for exterior lighting shall comply with the following:

1. They shall have a clock capable of being programmed for not fewer than 7 days.
2. They shall be capable of being set for seven different day types per week.
3. They shall incorporate an automatic holiday setback feature.
4. They shall have program backup capabilities that prevent the loss of program and time settings for a period of not less than 10 hours in the event that power is interrupted.

C405.3 Interior lighting power requirements (Prescriptive). A building complies with this section where its total connected interior lighting power calculated under Section

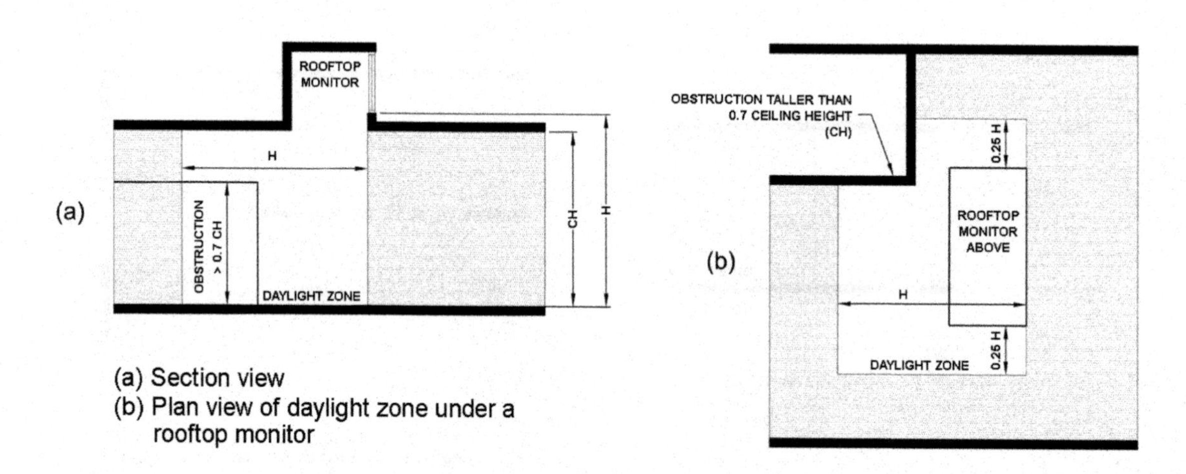

FIGURE C405.2.3.3(2)
DAYLIGHT ZONE UNDER A ROOFTOP MONITOR

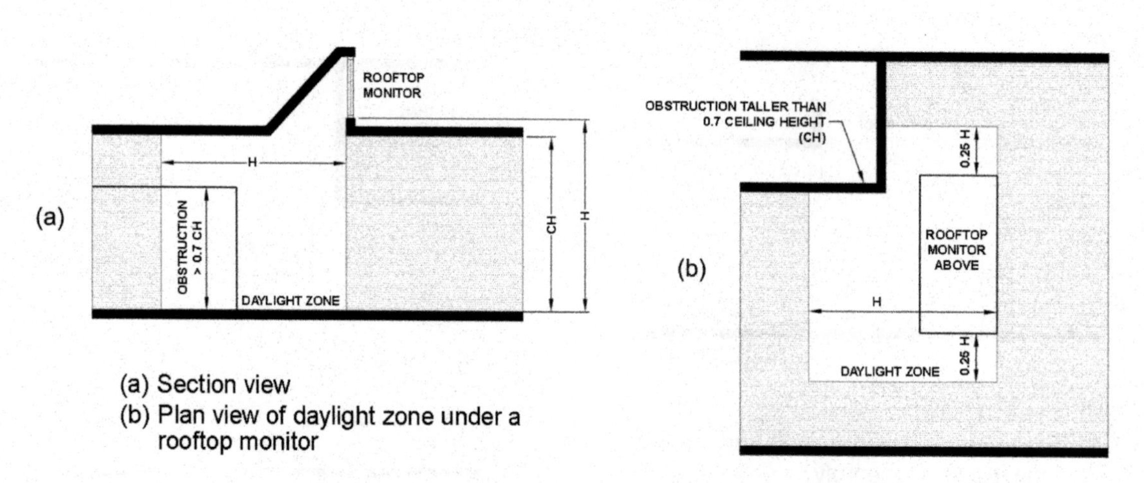

FIGURE C405.2.3.3(3)
DAYLIGHT ZONE UNDER A SLOPED ROOFTOP MONITOR

C405.3.1 is not greater than the interior lighting power allowance calculated under Section C405.3.2.

C405.3.1 Total connected interior lighting power. The total connected interior lighting power shall be determined in accordance with Equation 4-10.

$$TCLP = [LVL + BLL + LED + TRK + \text{Other}]$$ **(Equation 4-10)**

where:

TCLP = Total connected lighting power (watts).

LVL = For luminaires with lamps connected directly to building power, such as line voltage lamps, the rated wattage of the lamp.

BLL = For luminaires incorporating a ballast or transformer, the rated input wattage of the ballast or transformer when operating that lamp.

LED = For light-emitting diode luminaires with either integral or remote drivers, the rated wattage of the luminaire.

TRK = For lighting track, cable conductor, rail conductor, and plug-in busway systems that allow the addition and relocation of luminaires without rewiring, the wattage shall be one of the following:

1. The specified wattage of the luminaires, but not less than 8 W per linear foot (25 W/lin m).
2. The wattage limit of the permanent current-limiting devices protecting the system.
3. The wattage limit of the transformer supplying the system.

Other = The wattage of all other luminaires and lighting sources not covered previously and associated with interior lighting verified by data supplied by the manufacturer or other *approved* sources.

The connected power associated with the following lighting equipment and applications is not included in calculating total connected lighting power.

1. Television broadcast lighting for playing areas in sports arenas.
2. Emergency lighting automatically off during normal building operation.
3. Lighting in spaces specifically designed for use by occupants with special lighting needs, including those with visual impairment and other medical and age-related issues.
4. Casino gaming areas.
5. Mirror lighting in dressing rooms.
6. Task lighting for medical and dental purposes that is in addition to general lighting and controlled by an independent control device.
7. Display lighting for exhibits in galleries, museums and monuments that is in addition to general lighting and controlled by an independent control device.
8. Lighting for theatrical purposes, including performance, stage, film production and video production.
9. Lighting for photographic processes.
10. Lighting integral to equipment or instrumentation and installed by the manufacturer.
11. Task lighting for plant growth or maintenance.
12. Advertising signage or directional signage.
13. Lighting for food warming.
14. Lighting equipment that is for sale.
15. Lighting demonstration equipment in lighting education facilities.
16. Lighting approved because of safety considerations.
17. Lighting in retail display windows, provided that the display area is enclosed by ceiling-height partitions.
18. Furniture-mounted supplemental task lighting that is controlled by automatic shutoff.
19. Exit signs.

C405.3.2 Interior lighting power allowance. The total interior lighting power allowance (watts) is determined according to Table C405.3.2(1) using the Building Area Method, or Table C405.3.2(2) using the Space-by-Space Method, for all areas of the building covered in this permit.

C405.3.2.1 Building Area Method. For the Building Area Method, the interior lighting power allowance is the floor area for each building area type listed in Table C405.3.2(1) times the value from Table C405.3.2(1) for that area. For the purposes of this method, an "area" shall be defined as all contiguous spaces that accommodate or are associated with a single building area type, as listed in Table C405.3.2(1). Where this method is used to calculate the total interior lighting power for an entire building, each building area type shall be treated as a separate area.

C405.3.2.2 Space-by-Space Method. For the Space-by-Space Method, the interior lighting power allowance is determined by multiplying the floor area of each space times the value for the space type in Table C405.3.2(2) that most closely represents the proposed use of the space, and then summing the lighting power allowances for all spaces. Tradeoffs among spaces are permitted.

C405.3.2.2.1 Additional interior lighting power. Where using the Space-by-Space Method, an increase in the interior lighting power allowance is permitted for specific lighting functions. Additional power shall be permitted only where the specified lighting is installed and automatically controlled separately from the general lighting, to be turned off during nonbusiness hours. This additional power shall be used only for the specified luminaires and shall not be used for any other purpose. An increase

TABLE C405.3.2(1)
INTERIOR LIGHTING POWER ALLOWANCES:
BUILDING AREA METHOD

BUILDING AREA TYPE	LPD (w/ft²)
Automotive facility	0.71
Convention center	0.76
Courthouse	0.90
Dining: bar lounge/leisure	0.90
Dining: cafeteria/fast food	0.79
Dining: family	0.78
Dormitory[a, b]	0.61
Exercise center	0.65
Fire station[a]	0.53
Gymnasium	0.68
Health care clinic	0.82
Hospital[a]	1.05
Hotel/Motel[a, b]	0.75
Library	0.78
Manufacturing facility	0.90
Motion picture theater	0.83
Multifamily[c]	0.68
Museum	1.06
Office	0.79
Parking garage	0.15
Penitentiary	0.75
Performing arts theater	1.18
Police station	0.80
Post office	0.67
Religious building	0.94
Retail	1.06
School/university	0.81
Sports arena	0.87
Town hall	0.80
Transportation	0.61
Warehouse	0.48
Workshop	0.90

a. Where sleeping units are excluded from lighting power calculations by application of Section R404.1, neither the area of the sleeping units nor the wattage of lighting in the sleeping units is counted.

b. Where dwelling units are excluded from lighting power calculations by application of Section R404.1, neither the area of the dwelling units nor the wattage of lighting in the dwelling units is counted.

c. Dwelling units are excluded. Neither the area of the dwelling units nor the wattage of lighting in the dwelling units is counted.

TABLE C405.3.2(2)
INTERIOR LIGHTING POWER ALLOWANCES:
SPACE-BY-SPACE METHOD

COMMON SPACE TYPES[a]	LPD (watts/sq.ft)
Atrium	
Less than 40 feet in height	0.03 per foot in total height
Greater than 40 feet in height	0.40 + 0.02 per foot in total height
Audience seating area	
In an auditorium	0.63
In a convention center	0.82
In a gymnasium	0.65
In a motion picture theater	1.14
In a penitentiary	0.28
In a performing arts theater	2.03
In a religious building	1.53
In a sports arena	0.43
Otherwise	0.43
Banking activity area	0.86
Breakroom (See Lounge/breakroom)	
Classroom/lecture hall/training room	
In a penitentiary	1.34
Otherwise	0.96
Computer room	1.33
Conference/meeting/multipurpose room	1.07
Copy/print room	0.56
Corridor	
In a facility for the visually impaired (and not used primarily by the staff)[b]	0.92
In a hospital	0.92
In a manufacturing facility	0.29
Otherwise	0.66
Courtroom	1.39
Dining area	
In bar/lounge or leisure dining	0.93
In cafeteria or fast food dining	0.63
In a facility for the visually impaired (and not used primarily by the staff)[b]	2.00
In family dining	0.71
In a penitentiary	0.96
Otherwise	0.63
Electrical/mechanical room	0.43
Emergency vehicle garage	0.41

(continued)

**TABLE C405.3.2(2)—continued
INTERIOR LIGHTING POWER ALLOWANCES:
SPACE-BY-SPACE METHOD**

COMMON SPACE TYPES[a]	LPD (watts/sq.ft)
Food preparation area	1.06
Guestroom[c, d]	0.77
Laboratory	
In or as a classroom	1.20
Otherwise	1.45
Laundry/washing area	0.43
Loading dock, interior	0.58
Lobby	
For an elevator	0.68
In a facility for the visually impaired (and not used primarily by the staff)[b]	2.03
In a hotel	1.06
In a motion picture theater	0.45
In a performing arts theater	1.70
Otherwise	1.0
Locker room	0.48
Lounge/breakroom	
In a healthcare facility	0.78
Otherwise	0.62
Office	
Enclosed	0.93
Open plan	0.81
Parking area, interior	0.14
Pharmacy area	1.34
Restroom	
In a facility for the visually impaired (and not used primarily by the staff[b]	0.96
Otherwise	0.85
Sales area	1.22
Seating area, general	0.42
Stairway (see Space containing stairway)	
Stairwell	0.58
Storage room	0.46
Vehicular maintenance area	0.56
Workshop	1.14
BUILDING TYPE SPECIFIC SPACE TYPES[a]	**LPD (watts/sq.ft)**
Automotive (see Vehicular maintenance area)	
Convention Center—exhibit space	0.88
Dormitory—living quarters[c, d]	0.54
Facility for the visually impaired[b]	
In a chapel (and not used primarily by the staff)	1.06
In a recreation room (and not used primarily by the staff)	1.80
Fire Station—sleeping quarters[c]	0.20
Gymnasium/fitness center	
In an exercise area	0.50
In a playing area	0.82

(continued)

**TABLE C405.3.2(2)—continued
INTERIOR LIGHTING POWER ALLOWANCES:
SPACE-BY-SPACE METHOD**

BUILDING TYPE SPECIFIC SPACE TYPES[a]	LPD (watts/sq.ft)
Healthcare facility	
In an exam/treatment room	1.68
In an imaging room	1.06
In a medical supply room	0.54
In a nursery	1.00
In a nurse's station	0.81
In an operating room	2.17
In a patient room[c]	0.62
In a physical therapy room	0.84
In a recovery room	1.03
Library	
In a reading area	0.82
In the stacks	1.20
Manufacturing facility	
In a detailed manufacturing area	0.93
In an equipment room	0.65
In an extra-high-bay area (greater than 50′ floor-to-ceiling height)	1.05
In a high-bay area (25-50′ floor-to-ceiling height)	0.75
In a low-bay area (less than 25′ floor-to-ceiling height)	0.96
Museum	
In a general exhibition area	1.05
In a restoration room	0.85
Performing arts theater—dressing room	0.36
Post office—sorting area	0.68
Religious buildings	
In a fellowship hall	0.55
In a worship/pulpit/choir area	1.53
Retail facilities	
In a dressing/fitting room	0.50
In a mall concourse	0.90
Sports arena—playing area	
For a Class I facility[e]	2.47
For a Class II facility[f]	1.96
For a Class III facility[g]	1.70
For a Class IV facility[h]	1.13

(continued)

TABLE C405.3.2(2)—continued
INTERIOR LIGHTING POWER ALLOWANCES:
SPACE-BY-SPACE METHOD

BUILDING TYPE SPECIFIC SPACE TYPES[a]	LPD (watts/sq.ft)
Transportation facility	
In a baggage/carousel area	0.45
In an airport concourse	0.31
At a terminal ticket counter	0.62
Warehouse—storage area	
For medium to bulky, palletized items	0.35
For smaller, hand-carried items	0.69

a. In cases where both a common space type and a building area specific space type are listed, the building area specific space type shall apply
b. A 'Facility for the Visually Impaired' is a facility that is licensed or will be licensed by local or state authorities for senior long-term care, adult daycare, senior support or people with special visual needs.
c. Where sleeping units are excluded from lighting power calculations by application of Section R404.1, neither the area of the sleeping units nor the wattage of lighting in the sleeping units is counted.
d. Where dwelling units are excluded from lighting power calculations by application of Section R404.1, neither the area of the dwelling units nor the wattage of lighting in the dwelling units is counted.
e. Class I facilities consist of professional facilities; and semiprofessional, collegiate, or club facilities with seating for 5,000 or more spectators.
f. Class II facilities consist of collegiate and semiprofessional facilities with seating for fewer than 5,000 spectators; club facilities with seating for between 2,000 and 5,000 spectators; and amateur league and high-school facilities with seating for more than 2,000 spectators.
g. Class III facilities consist of club, amateur league and high-school facilities with seating for 2,000 or fewer spectators.
h. Class IV facilities consist of elementary school and recreational facilities; and amateur league and high-school facilities without provision for spectators.

in the interior lighting power allowance is permitted in the following cases:

1. For lighting equipment to be installed in sales areas specifically to highlight merchandise, the additional lighting power shall be determined in accordance with Equation 4-11.

   Additional interior lighting power allowance = 1000 W + (Retail Area 1 × 0.45 W/ft^2) + (Retail Area 2 × 0.45W/ft^2) + (Retail Area 3 × 1.05 W/ft^2) + (Retail Area 4 × 1.87 W/ft^2)

   For SI units:

   Additional interior lighting power allowance = 1000 W + (Retail Area 1 × 4.8 W/m^2) + (Retail Area 2 × 4.84 W/m^2) + (Retail Area 3 × 11 W/m^2) + (Retail Area 4 × 20 W/m^2)

   (Equation 4-11)

   where:

   Retail Area 1 = The floor area for all products not listed in Retail Area 2, 3 or 4.

   Retail Area 2 = The floor area used for the sale of vehicles, sporting goods and small electronics.

   Retail Area 3 = The floor area used for the sale of furniture, clothing, cosmetics and artwork.

   Retail Area 4 = The floor area used for the sale of jewelry, crystal and china.

   Exception: Other merchandise categories are permitted to be included in Retail Areas 2 through 4, provided that justification documenting the need for additional lighting power based on visual inspection, contrast, or other critical display is approved by the code official.

2. For spaces in which lighting is specified to be installed in addition to the general lighting for the purpose of decorative appearance or for highlighting art or exhibits, provided that the additional lighting power shall be not more than 0.9 W/ft^2 (9.7 W/m^2) in lobbies and not more than 0.75 W/ft^2 (8.1 W/m^2) in other spaces.

C405.4 Exterior lighting power requirements (Mandatory). The total connected exterior lighting power calculated in accordance with Section C405.4.1 shall be not greater than the exterior lighting power allowance calculated in accordance with Section C405.4.2.

C405.4.1 Total connected exterior building exterior lighting power. The total exterior connected lighting power shall be the total maximum rated wattage of all lighting that is powered through the energy service for the building.

Exception: Lighting used for the following applications shall not be included.

1. Lighting *approved* because of safety considerations.
2. Emergency lighting automatically off during normal business operation.
3. Exit signs.
4. Specialized signal, directional and marker lighting associated with transportation.
5. Advertising signage or directional signage.
6. Integral to equipment or instrumentation and installed by its manufacturer.
7. Theatrical purposes, including performance, stage, film production and video production.
8. Athletic playing areas.
9. Temporary lighting.
10. Industrial production, material handling, transportation sites and associated storage areas.
11. Theme elements in theme/amusement parks.
12. Used to highlight features of art, public monuments, and the national flag.
13. Lighting for water features and swimming pools.
14. Lighting controlled from within dwelling units, where the lighting complies with Section R404.1.

C405.4.2 Exterior lighting power allowance. The total exterior lighting power allowance is the sum of the base site allowance plus the individual allowances for areas that are to be illuminated by lighting that is powered through the energy service for the building. Lighting power allowances are as specified in Table C405.4.2(2). The lighting zone for the building exterior is determined in accordance with Table C405.4.2(1) unless otherwise specified by the *code official*.

**TABLE C405.4.2(1)
EXTERIOR LIGHTING ZONES**

LIGHTING ZONE	DESCRIPTION
1	Developed areas of national parks, state parks, forest land, and rural areas
2	Areas predominantly consisting of residential zoning, neighborhood business districts, light industrial with limited nighttime use and residential mixed-use areas
3	All other areas not classified as lighting zone 1, 2 or 4
4	High-activity commercial districts in major metropolitan areas as designated by the local land use planning authority

C405.4.2.1 Additional exterior lighting power. Any increase in the exterior lighting power allowance is limited to the specific lighting applications indicated in Table C405.4.2(3). The additional power shall be used only for the luminaires that are serving these applications and shall not be used for any other purpose.

C405.4.3 Gas lighting (Mandatory). Gas-fired lighting appliances shall not be equipped with continuously burning pilot ignition systems.

C405.5 Dwelling electrical meter (Mandatory). Each dwelling unit located in a *Group R-2* building shall have a separate electrical meter.

C405.6 Electrical transformers (Mandatory). Low-voltage dry-type distribution electric transformers shall meet the minimum efficiency requirements of Table C405.6 as tested and rated in accordance with the test procedure listed in DOE 10 CFR 431. The efficiency shall be verified through certification under an approved certification program or, where a certification program does not exist, the equipment efficiency ratings shall be supported by data furnished by the transformer manufacturer.

Exceptions: The following transformers are exempt:

1. Transformers that meet the *Energy Policy Act of 2005* exclusions based on the DOE 10 CFR 431 definition of special purpose applications.
2. Transformers that meet the *Energy Policy Act of 2005* exclusions that are not to be used in general purpose applications based on information provided in DOE 10 CFR 431.
3. Transformers that meet the *Energy Policy Act of 2005* exclusions with multiple voltage taps where the highest tap is not less than 20 percent more than the lowest tap.
4. Drive transformers.
5. Rectifier transformers.
6. Auto-transformers.
7. Uninterruptible power system transformers.
8. Impedance transformers.
9. Regulating transformers.
10. Sealed and nonventilating transformers.
11. Machine tool transformers.
12. Welding transformers.
13. Grounding transformers.
14. Testing transformers.

C405.7 Electric motors (Mandatory). Electric motors shall meet the minimum efficiency requirements of Tables C405.7(1) through C405.7(4) when tested and rated in accordance with the DOE 10 CFR 431. The efficiency shall be verified through certification under an approved certification program or, where a certification program does not exist, the equipment efficiency ratings shall be supported by data furnished by the motor manufacturer.

Exception: The standards in this section shall not apply to the following exempt electric motors:

1. Air-over electric motors.
2. Component sets of an electric motor.
3. Liquid-cooled electric motors.
4. Submersible electric motors.
5. Inverter-only electric motors.

C405.8 Vertical and horizontal transportation systems and equipment. Vertical and horizontal transportation systems and equipment shall comply with this section.

C405.8.1 Elevator cabs. For the luminaires in each elevator cab, not including signals and displays, the sum of the lumens divided by the sum of the watts shall be not less than 35 lumens per watt. Ventilation fans in elevators that do not have their own air-conditioning system shall not consume more than 0.33 watts/cfm at the maximum rated speed of the fan. Controls shall be provided that will de-energize ventilation fans and lighting systems when the elevator is stopped, unoccupied and with its doors closed for over 15 minutes.

C405.8.2 Escalators and moving walks. Escalators and moving walks shall comply with ASME A17.1/CSA B44 and shall have automatic controls configured to reduce speed to the minimum permitted speed in accordance with ASME A17.1/CSA B44 or applicable local code when not conveying passengers.

Exception: A variable voltage drive system that reduces operating voltage in response to light loading conditions is an alternative to the reduced speed function.

C405.8.2.1 Regenerative drive. An escalator designed either for one-way down operation only or for reversible operation shall have a variable frequency regenerative drive that supplies electrical energy to the building electrical system when the escalator is loaded with passengers whose combined weight exceeds 750 pounds (340 kg).

TABLE C405.4.2(2)
LIGHTING POWER ALLOWANCES FOR BUILDING EXTERIORS

	LIGHTING ZONES			
	Zone 1	Zone 2	Zone 3	Zone 4
Base Site Allowance	350 W	400 W	500 W	900 W
Uncovered Parking Areas				
Parking areas and drives	0.03W/ft^2	0.04 W/ft^2	0.06 W/ft^2	0.08 W/ft^2
Building Grounds				
Walkways and ramps less than 10 feet wide	0.5 W/linear foot	0.5 W/linear foot	0.6 W/linear foot	0.7 W/linear foot
Walkways and ramps 10 feet wide or greater, plaza areas, special feature areas	0.10 W/ft^2	0.10 W/ft^2	0.11 W/ft^2	0.14 W/ft^2
Dining areas	0.65 W/ft^2	0.65 W/ft^2	0.75 W/ft^2	0.95 W/ft^2
Stairways	0.6 W/ft^2	0.7 W/ft^2	0.7 W/ft^2	0.7 W/ft^2
Pedestrian tunnels	0.12 W/ft^2	0.12 W/ft^2	0.14 W/ft^2	0.21 W/ft^2
Landscaping	0.03 W/ft^2	0.04 W/ft^2	0.04 W/ft^2	0.04 W/ft^2
Building Entrances and Exits				
Pedestrian and vehicular entrances and exits	14 W/linear foot of opening	14 W/linear foot of opening	21 W/linear foot of opening	21 W/linear foot of opening
Entry canopies	0.20 W/ft^2	0.25 W/ft^2	0.4 W/ft^2	0.4 W/ft^2
Loading docks	0.35 W/ft^2	0.35 W/ft^2	0.35 W/ft^2	0.35 W/ft^2
Sales Canopies				
Free-standing and attached	0.40 W/ft^2	0.40 W/ft^2	0.6 W/ft^2	0.7 W/ft^2
Outdoor Sales				
Open areas (including vehicle sales lots)	0.20 W/ft^2	0.20 W/ft^2	0.35 W/ft^2	0.50 W/ft^2
Street frontage for vehicle sales lots in addition to "open area" allowance	No allowance	7 W/linear foot	7 W/linear foot	21 W/linear foot

For SI: 1 foot = 304.8 mm, 1 watt per square foot = W/0.0929 m^2.
W = watts.

TABLE C405.4.2(3)
INDIVIDUAL LIGHTING POWER ALLOWANCES FOR BUILDING EXTERIORS

	LIGHTING ZONES			
	Zone 1	Zone 2	Zone 3	Zone 4
Building facades	No allowance	0.075 W/ft^2 of gross above-grade wall area	0.113 W/ft^2 of gross above-grade wall area	0.15 W/ft^2 of gross above-grade wall area
Automated teller machines (ATM) and night depositories	135 W per location plus 45 W per additional ATM per location			
Uncovered entrances and gatehouse inspection stations at guarded facilities	0.5 W/ft^2 of area			
Uncovered loading areas for law enforcement, fire, ambulance and other emergency service vehicles	0.35 W/ft^2 of area			
Drive-up windows and doors	200 W per drive through			
Parking near 24-hour retail entrances.	400 W per main entry			

For SI: 1 watt per square foot = W/0.0929 m^2.
W = watts.

TABLE C405.6
MINIMUM NOMINAL EFFICIENCY LEVELS FOR 10 CFR 431 LOW-VOLTAGE DRY-TYPE DISTRIBUTION TRANSFORMERS

SINGLE-PHASE TRANSFORMERS		THREE-PHASE TRANSFORMERS	
kVA[a]	Efficiency (%)[b]	kVA[a]	Efficiency (%)[b]
15	97.70	15	97.89
25	98.00	30	98.23
37.5	98.20	45	98.40
50	98.30	75	98.60
75	98.50	112.5	98.74
100	98.60	150	98.83
167	98.70	225	98.94
250	98.80	300	99.02
333	98.90	500	99.14
—	—	750	99.23
—	—	1000	99.28

a. kiloVolt-Amp rating.
b. Nominal efficiencies shall be established in accordance with the DOE 10 CFR 431 test procedure for low-voltage dry-type transformers.

C405.9 Voltage drop in feeders and branch circuits. The total *voltage drop* across the combination of feeders and branch circuits shall not exceed 5 percent.

SECTION C406
ADDITIONAL EFFICIENCY PACKAGES

C406.1 Requirements. Buildings shall comply with one or more of the following:

1. More efficient HVAC performance in accordance with Section C406.2.
2. Reduced lighting power in accordance with Section C406.3.
3. Enhanced lighting controls in accordance with Section C406.4.
4. On-site supply of renewable energy in accordance with Section C406.5.
5. Provision of a dedicated outdoor air system for certain HVAC equipment in accordance with Section C406.6.
6. High-efficiency service water heating in accordance with Section C406.7.
7. Enhanced envelope performance in accordance with Section C406.8.
8. Reduced air infiltration in accordance with Section C406.9

C406.1.1 Tenant spaces. Tenant spaces shall comply with Section C406.2, C406.3, C406.4, C406.6 or C406.7. Alternatively, tenant spaces shall comply with Section C406.5 where the entire building is in compliance.

Exception: Previously occupied tenant spaces that comply with this code in accordance with Section C501.

C406.2 More efficient HVAC equipment performance. Equipment shall exceed the minimum efficiency requirements listed in Tables C403.3.2(1) through C403.3.2(7) by 10 percent, in addition to the requirements of Section C403. Where multiple performance requirements are provided, the equipment shall exceed all requirements by 10 percent. *Variable refrigerant flow systems* shall exceed the energy efficiency provisions of ANSI/ASHRAE/IESNA 90.1 by 10 percent. Equipment not listed in Tables C403.3.2(1) through C403.3.2(7) shall be limited to 10 percent of the total building system capacity.

C406.3 Reduced lighting power. The total connected interior lighting power calculated in accordance with Section C405.3.1 shall be less than 90 percent of the total lighting power allowance calculated in accordance with Section C405.3.2.

C406.4 Enhanced digital lighting controls. Interior lighting in the building shall have the following enhanced lighting controls that shall be located, scheduled and operated in accordance with Sections C405.2.1 through C405.2.3.

1. Luminaires shall be configured for continuous dimming.
2. Luminaires shall be addressed individually. Where individual addressability is not available for the luminaire class type, a controlled group of not more than four luminaries shall be allowed.
3. Not more than eight luminaires shall be controlled together in a *daylight zone.*
4. Fixtures shall be controlled through a digital control system that includes the following function:
   - 4.1. Control reconfiguration based on digital addressability.
   - 4.2. Load shedding.
   - 4.3. Individual user control of overhead general illumination in open offices.
   - 4.4. Occupancy sensors shall be capable of being reconfigured through the digital control system.

TABLE C405.7(1)
MINIMUM NOMINAL FULL-LOAD EFFICIENCY FOR NEMA DESIGN A, NEMA DESIGN B, AND *IEC DESIGN N MOTORS* (EXCLUDING FIRE PUMP) ELECTRIC MOTORS AT 60 HZ[a, b]

MOTOR HORSEPOWER (STANDARD KILOWATT EQUIVALENT)	NOMINAL FULL-LOAD EFFICIENCY (%) AS OF JUNE 1, 2016							
	2 Pole		4 Pole		6 Pole		8 Pole	
	Enclosed	Open	Enclosed	Open	Enclosed	Open	Enclosed	Open
1 (0.75)	77.0	77.0	85.5	85.5	82.5	82.5	75.5	75.5
1.5 (1.1)	84.0	84.0	86.5	86.5	87.5	86.5	78.5	77.0
2 (1.5)	85.5	85.5	86.5	86.5	88.5	87.5	84.0	86.5
3 (2.2)	86.5	85.5	89.5	89.5	89.5	88.5	85.5	87.5
5 (3.7)	88.5	86.5	89.5	89.5	89.5	89.5	86.5	88.5
7.5 (5.5)	89.5	88.5	91.7	91.0	91.0	90.2	86.5	89.5
10 (7.5)	90.2	89.5	91.7	91.7	91.0	91.7	89.5	90.2
15 (11)	91.0	90.2	92.4	93.0	91.7	91.7	89.5	90.2
20 (15)	91.0	91.0	93.0	93.0	91.7	92.4	90.2	91.0
25 (18.5)	91.7	91.7	93.6	93.6	93.0	93.0	90.2	91.0
30 (22)	91.7	91.7	93.6	94.1	93.0	93.6	91.7	91.7
40 (30)	92.4	92.4	94.1	94.1	94.1	94.1	91.7	91.7
50 (37)	93.0	93.0	94.5	94.5	94.1	94.1	92.4	92.4
60 (45)	93.6	93.6	95.0	95.0	94.5	94.5	92.4	93.0
75 (55)	93.6	93.6	95.4	95.0	94.5	94.5	93.6	94.1
100 (75)	94.1	93.6	95.4	95.4	95.0	95.0	93.6	94.1
125 (90)	95.0	94.1	95.4	95.4	95.0	95.0	94.1	94.1
150 (110)	95.0	94.1	95.8	95.8	95.8	95.4	94.1	94.1
200 (150)	95.4	95.0	96.2	95.8	95.8	95.4	94.5	94.1
250 (186)	95.8	95.0	96.2	95.8	95.8	95.8	95.0	95.0
300 (224)	95.8	95.4	96.2	95.8	95.8	95.8		
350 (261)	95.8	95.4	96.2	95.8	95.8	95.8		
400 (298)	95.8	95.8	96.2	95.8				
450 (336)	95.8	96.2	96.2	96.2				
500 (373)	95.8	96.2	96.2	96.2				

a. Nominal efficiencies shall be established in accordance with DOE 10 CFR 431.

b. For purposes of determining the required minimum nominal full-load efficiency of an electric motor that has a horsepower or kilowatt rating between two horsepower or two kilowatt ratings listed in this table, each such motor shall be deemed to have a listed horsepower or kilowatt rating, determined as follows:
   1. A horsepower at or above the midpoint between the two consecutive horsepowers shall be rounded up to the higher of the two horsepowers.
   2. A horsepower below the midpoint between the two consecutive horsepowers shall be rounded down to the lower of the two horsepowers.
   3. A kilowatt rating shall be directly converted from kilowatts to horsepower using the formula: 1 kilowatt = (1/0.746) horsepower. The conversion should be calculated to three significant decimal places, and the resulting horsepower shall be rounded in accordance with No. 1 or No. 2 above, as applicable.

TABLE C405.7(2)
MINIMUM NOMINAL FULL-LOAD EFFICIENCY FOR NEMA DESIGN C AND *IEC DESIGN H MOTORS* AT 60 HZ[a, b]

MOTOR HORSEPOWER (STANDARD KILOWATT EQUIVALENT)	NOMINAL FULL-LOAD EFFICIENCY (%) AS OF JUNE 1, 2016					
	4 Pole		6 Pole		8 Pole	
	Enclosed	Open	Enclosed	Open	Enclosed	Open
1 (0.75)	85.5	85.5	82.5	82.5	75.5	75.5
1.5 (1.1)	86.5	86.5	87.5	86.5	78.5	77.0
2 (1.5)	86.5	86.5	88.5	87.5	84.0	86.5
3 (2.2)	89.5	89.5	89.5	88.5	85.5	87.5
5 (3.7)	89.5	89.5	89.5	89.5	86.5	88.5
7.5 (5.5)	91.7	91.0	91.0	90.2	86.5	89.5
10 (7.5)	91.7	91.7	91.0	91.7	89.5	90.2
15 (11)	92.4	93.0	91.7	91.7	89.5	90.2
20 (15)	93.0	93.0	91.7	92.4	90.2	91.0
25 (18.5)	93.6	93.6	93.0	93.0	90.2	91.0
30 (22)	93.6	94.1	93.0	93.6	91.7	91.7
40 (30)	94.1	94.1	94.1	94.1	91.7	91.7
50 (37)	94.5	94.5	94.1	94.1	92.4	92.4
60 (45)	95.0	95.0	94.5	94.5	92.4	93.0
75 (55)	95.4	95.0	94.5	94.5	93.6	94.1
100 (75)	95.4	95.4	95.0	95.0	93.6	94.1
125 (90)	95.4	95.4	95.0	95.0	94.1	94.1
150 (110)	95.8	95.8	95.8	95.4	94.1	94.1
200 (150)	96.2	95.8	95.8	95.4	94.5	94.1

a. Nominal efficiencies shall be established in accordance with DOE 10 CFR 431.

b. For purposes of determining the required minimum nominal full-load efficiency of an electric motor that has a horsepower or kilowatt rating between two horsepower or two kilowatt ratings listed in this table, each such motor shall be deemed to have a listed horsepower or kilowatt rating, determined as follows:
1. A horsepower at or above the midpoint between the two consecutive horsepowers shall be rounded up to the higher of the two horsepowers.
2. A horsepower below the midpoint between the two consecutive horsepowers shall be rounded down to the lower of the two horsepowers.
3. A kilowatt rating shall be directly converted from kilowatts to horsepower using the formula: 1 kilowatt = (1/0.746) horsepower. The conversion should be calculated to three significant decimal places, and the resulting horsepower shall be rounded in accordance with No. 1 or No. 2 above, as applicable.

TABLE C405.7(3)
MINIMUM AVERAGE FULL-LOAD EFFICIENCY POLYPHASE SMALL ELECTRIC MOTORS[a]

MOTOR HORSEPOWER	OPEN MOTORS			
	Number of Poles	2	4	6
	Synchronous Speed (RPM)	3600	1800	1200
0.25		65.6	69.5	67.5
0.33		69.5	73.4	71.4
0.50		73.4	78.2	75.3
0.75		76.8	81.1	81.7
1		77.0	83.5	82.5
1.5		84.0	86.5	83.8
2		85.5	86.5	N/A
3		85.5	86.9	N/A

a. Average full-load efficiencies shall be established in accordance with DOE 10 CFR 431.

TABLE C405.7(4)
MINIMUM AVERAGE FULL-LOAD EFFICIENCY FOR CAPACITOR-START CAPACITOR-RUN AND CAPACITOR-START INDUCTION-RUN SMALL ELECTRIC MOTORS[a]

MOTOR HORSEPOWER	OPEN MOTORS			
	Number of Poles	2	4	6
	Synchronous Speed (RPM)	3600	1800	1200
0.25		66.6	68.5	62.2
0.33		70.5	72.4	66.6
0.50		72.4	76.2	76.2
0.75		76.2	81.8	80.2
1		80.4	82.6	81.1
1.5		81.5	83.8	N/A
2		82.9	84.5	N/A
3		84.1	N/A	N/A

a. Average full-load efficiencies shall be established in accordance with DOE 10 CFR 431.

5. Construction documents shall include submittal of a Sequence of Operations, including a specification outlining each of the functions in Item 4.
6. Functional testing of lighting controls shall comply with Section C408.

C406.5 On-site renewable energy. The total minimum ratings of on-site renewable energy systems shall be one of the following:

1. Not less than 1.71 Btu/h per square foot (5.4 W/m^2) or 0.50 watts per square foot (5.4 W/m^2) of conditioned floor area.
2. Not less than 3 percent of the energy used within the building for building mechanical and service water heating equipment and lighting regulated in Chapter 4.

C406.6 Dedicated outdoor air system. Buildings containing equipment or systems regulated by Section C403.3.4, C403.4.3, C403.4.4, C403.4.5, C403.6, C403.8.4, C403.8.5, C403.8.5.1, C403.9.1, C403.9.2, C403.9.3 or C403.9.4 shall be equipped with an independent ventilation system designed to provide not less than the minimum 100-percent outdoor air to each individual occupied space, as specified by the *International Mechanical Code.* The ventilation system shall be capable of total energy recovery. The HVAC system shall include supply-air temperature controls that automatically reset the supply-air temperature in response to representative building loads, or to outdoor air temperatures. The controls shall reset the supply-air temperature not less than 25 percent of the difference between the design supply-air temperature and the design room-air temperature.

C406.7 Reduced energy use in service water heating. Buildings shall be of the following types to use this compliance method:

1. *Group R*-1: Boarding houses, hotels or motels.
2. Group I-2: Hospitals, psychiatric hospitals and nursing homes.
3. Group A-2: Restaurants and banquet halls or buildings containing food preparation areas.
4. Group F: Laundries.
5. *Group R*-2.
6. Group A-3: Health clubs and spas.
7. Buildings showing a service hot water load of 10 percent or more of total building energy loads, as shown with an energy analysis as described in Section C407.

C406.7.1 Load fraction. The building service water-heating system shall have one or more of the following that are sized to provide not less than 60 percent of the building's annual hot water requirements, or sized to provide 100 percent of the building's annual hot water requirements if the building shall otherwise comply with Section C403.9.5:

1. Waste heat recovery from service hot water, heat-recovery chillers, building equipment, or process equipment.
2. *On-site renewable energy* water-heating systems.

C406.8 Enhanced envelope performance. The total UA of the building thermal envelope as designed shall be not less than 15 percent below the total UA of the building thermal envelope in accordance with Section C402.1.5.

C406.9 Reduced air infiltration. Air infiltration shall be verified by whole-building pressurization testing conducted in accordance with ASTM E779 or ASTM E1827 by an independent third party. The measured air-leakage rate of the building envelope shall not exceed 0.25 cfm/ft^2 (2.0 L/s × m^2) under a pressure differential of 0.3 inches water column (75 Pa), with the calculated surface area being the sum of the above- and below-grade building envelope. A report that includes the tested surface area, floor area, air by volume, stories above grade, and leakage rates shall be submitted to the code official and the building owner.

Exception: For buildings having over 250,000 square feet (25 000 m^2) of conditioned floor area, air leakage testing need not be conducted on the whole building where testing is conducted on representative above-grade sections of the building. Tested areas shall total not less than 25 percent of the conditioned floor area and shall be tested in accordance with this section.

SECTION C407
TOTAL BUILDING PERFORMANCE

C407.1 Scope. This section establishes criteria for compliance using total building performance. The following systems and loads shall be included in determining the total building performance: heating systems, cooling systems, service water heating, fan systems, lighting power, receptacle loads and process loads.

Exception: Energy used to recharge or refuel vehicles that are used for on-road and off-site transportation purposes.

C407.2 Mandatory requirements. Compliance with this section requires compliance with Sections C402.5, C403.2, C403.3 through C403.3.2, C403.4 through C403.4.2.3, C403.5.5, C403.7, C403.8.1 through C403.8.4, C403.10.1 through C403.10.3, C403.11, C403.12, C404 and C405.

C407.3 Performance-based compliance. Compliance based on total building performance requires that a proposed building (*proposed design*) be shown to have an annual energy cost that is less than or equal to the annual energy cost of the *standard reference design*. Energy prices shall be taken from a source *approved* by the *code official*, such as the Department of Energy, Energy Information Administration's *State Energy Price and Expenditure Report*. *Code officials* shall be permitted to require time-of-use pricing in energy cost calculations. The reduction in energy cost of the proposed design associated with *on-site renewable energy* shall be not more than 5 percent of the total energy cost. The amount of renewable energy purchased from off-site sources shall be the same in the *standard reference design* and the *proposed design*.

Exception: Jurisdictions that require site energy (1 kWh = 3413 Btu) rather than energy cost as the metric of comparison.

C407.4 Documentation. Documentation verifying that the methods and accuracy of compliance software tools conform to the provisions of this section shall be provided to the *code official*.

C407.4.1 Compliance report. Permit submittals shall include a report documenting that the proposed design has annual energy costs less than or equal to the annual energy costs of the standard reference design. The compliance documentation shall include the following information:

1. Address of the building.
2. An inspection checklist documenting the building component characteristics of the *proposed design* as specified in Table C407.5.1(1). The inspection checklist shall show the estimated annual energy cost for both the *standard reference design* and the *proposed design*.
3. Name of individual completing the compliance report.
4. Name and version of the compliance software tool.

C407.4.2 Additional documentation. The *code official* shall be permitted to require the following documents:

1. Documentation of the building component characteristics of the *standard reference design*.
2. Thermal zoning diagrams consisting of floor plans showing the thermal zoning scheme for *standard reference design* and *proposed design*.
3. Input and output reports from the energy analysis simulation program containing the complete input and output files, as applicable. The output file shall include energy use totals and energy use by energy source and end-use served, total hours that space conditioning loads are not met and any errors or warning messages generated by the simulation tool as applicable.
4. An explanation of any error or warning messages appearing in the simulation tool output.
5. A certification signed by the builder providing the building component characteristics of the *proposed design* as given in Table C407.5.1(1).
6. Documentation of the reduction in energy use associated with *on-site renewable energy*.

C407.5 Calculation procedure. Except as specified by this section, the *standard reference design* and *proposed design* shall be configured and analyzed using identical methods and techniques.

C407.5.1 Building specifications. The *standard reference design* and *proposed design* shall be configured and analyzed as specified by Table C407.5.1(1). Table C407.5.1(1) shall include by reference all notes contained in Table C402.1.4.

C407.5.2 Thermal blocks. The *standard reference design* and *proposed design* shall be analyzed using identical thermal blocks as specified in Section C407.5.2.1, C407.5.2.2 or C407.5.2.3.

C407.5.2.1 HVAC zones designed. Where HVAC *zones* are defined on HVAC design drawings, each HVAC *zone* shall be modeled as a separate thermal block.

Exception: Different HVAC *zones* shall be allowed to be combined to create a single thermal block or identical thermal blocks to which multipliers are applied, provided that:

1. The space use classification is the same throughout the thermal block.
2. All HVAC *zones* in the thermal block that are adjacent to glazed exterior walls face the same orientation or their orientations are within 45 degrees (0.79 rad) of each other.
3. All of the *zones* are served by the same HVAC system or by the same kind of HVAC system.

C407.5.2.2 HVAC zones not designed. Where HVAC *zones* have not yet been designed, thermal blocks shall be defined based on similar internal load densities, occupancy, lighting, thermal and temperature schedules, and in combination with the following guidelines:

1. Separate thermal blocks shall be assumed for interior and perimeter spaces. Interior spaces shall be those located more than 15 feet (4572 mm) from an exterior wall. Perimeter spaces shall be those located closer than 15 feet (4572 mm) from an *exterior wall*.
2. Separate thermal blocks shall be assumed for spaces adjacent to glazed exterior walls: a separate *zone* shall be provided for each orientation, except orientations that differ by not more than 45 degrees (0.79 rad) shall be permitted to be considered to be the same orientation. Each *zone* shall include floor area that is 15 feet (4572 mm) or less from a glazed perimeter wall, except that floor area within 15 feet (4572 mm) of glazed perimeter walls having more than one orientation shall be divided proportionately between *zones*.
3. Separate thermal blocks shall be assumed for spaces having floors that are in contact with the ground or exposed to ambient conditions from *zones* that do not share these features.
4. Separate thermal blocks shall be assumed for spaces having exterior ceiling or roof assemblies from *zones* that do not share these features.

C407.5.2.3 *Group R*-2 occupancy buildings. *Group R*-2 occupancy spaces shall be modeled using one thermal block per space except that those facing the same orientations are permitted to be combined into one thermal block. Corner units and units with roof or floor loads shall only be combined with units sharing these features.

C407.6 Calculation software tools. Calculation procedures used to comply with this section shall be software tools capable of calculating the annual energy consumption of all building elements that differ between the *standard reference design* and the *proposed design* and shall include the following capabilities.

1. Building operation for a full calendar year (8,760 hours).
2. Climate data for a full calendar year (8,760 hours) and shall reflect *approved* coincident hourly data for temperature, solar radiation, humidity and wind speed for the building location.
3. Ten or more thermal zones.
4. Thermal mass effects.
5. Hourly variations in occupancy, illumination, receptacle loads, thermostat settings, mechanical ventilation, HVAC equipment availability, service hot water usage and any process loads.
6. Part-load performance curves for mechanical equipment.
7. Capacity and efficiency correction curves for mechanical heating and cooling equipment.
8. Printed *code official* inspection checklist listing each of the *proposed design* component characteristics from Table C407.5.1(1) determined by the analysis to provide compliance, along with their respective performance ratings including, but not limited to, *R*-value, *U*-factor, SHGC, HSPF, AFUE, SEER, EF.

C407.6.1 Specific approval. Performance analysis tools complying with the applicable subsections of Section C407 and tested according to ASHRAE Standard 140 shall be permitted to be *approved*. Tools are permitted to be *approved* based on meeting a specified threshold for a jurisdiction. The *code official* shall be permitted to approve tools for a specified application or limited scope.

C407.6.2 Input values. Where calculations require input values not specified by Sections C402, C403, C404 and C405, those input values shall be taken from an *approved* source.

C407.6.3 Exceptional calculation methods. Where the simulation program does not model a design, material or device of the *proposed design*, an exceptional calculation method shall be used where approved by the *code official*. Where there are multiple designs, materials or devices that the simulation program does not model, each shall be calculated separately and exceptional savings determined for each. The total exceptional savings shall not constitute more than half of the difference between the baseline building performance and the proposed building performance. Applications for approval of an exceptional method shall include all of the following:

1. Step-by-step documentation of the exceptional calculation method performed, detailed enough to reproduce the results.
2. Copies of all spreadsheets used to perform the calculations.
3. A sensitivity analysis of energy consumption where each of the input parameters is varied from half to double the value assumed.
4. The calculations shall be performed on a time step basis consistent with the simulation program used.
5. The performance rating calculated with and without the exceptional calculation method.

TABLE C407.5.1(1)
SPECIFICATIONS FOR THE STANDARD REFERENCE AND PROPOSED DESIGNS

BUILDING COMPONENT CHARACTERISTICS	STANDARD REFERENCE DESIGN	PROPOSED DESIGN
Space use classification	Same as proposed	The space use classification shall be chosen in accordance with Table C405.5.2 for all areas of the building covered by this permit. Where the space use classification for a building is not known, the building shall be categorized as an office building.
Roofs	Type: Insulation entirely above deck	As proposed
	Gross area: same as proposed	As proposed
	U-factor: as specified in Table C402.1.4	As proposed
	Solar absorptance: 0.75	As proposed
	Emittance: 0.90	As proposed
Walls, above-grade	Type: Mass wall where proposed wall is mass; otherwise steel-framed wall	As proposed
	Gross area: same as proposed	As proposed
	U-factor: as specified in Table C402.1.4	As proposed
	Solar absorptance: 0.75	As proposed
	Emittance: 0.90	As proposed
Walls, below-grade	Type: Mass wall	As proposed
	Gross area: same as proposed	As proposed
	U-Factor: as specified in Table C402.1.4 with insulation layer on interior side of walls	As proposed
Floors, above-grade	Type: joist/framed floor	As proposed
	Gross area: same as proposed	As proposed
	U-factor: as specified in Table C402.1.4	As proposed
Floors, slab-on-grade	Type: Unheated	As proposed
	F-factor: as specified in Table C402.1.4	As proposed
Opaque doors	Type: Swinging	As proposed
	Area: Same as proposed	As proposed
	U-factor: as specified in Table C402.1.4	As proposed
Vertical fenestration other than opaque doors	Area 1. The proposed vertical fenestration area; where the proposed vertical fenestration area is less than 40 percent of above-grade wall area. 2. 40 percent of above-grade wall area; where the proposed vertical fenestration area is 40 percent or more of the above-grade wall area.	As proposed
	U-factor: as specified in Table C402.4	As proposed
	SHGC: as specified in Table C402.4 except that for climates with no requirement (NR) SHGC = 0.40 shall be used	As proposed
	External shading and PF: None	As proposed
Skylights	Area 1. The proposed skylight area; where the proposed skylight area is less than that permitted by Section C402.1. 2. The area permitted by Section C402.1; where the proposed skylight area exceeds that permitted by Section C402.1	As proposed
	U-factor: as specified in Table C402.4	As proposed
	SHGC: as specified in Table C402.4 except that for climates with no requirement (NR) SHGC = 0.40 shall be used.	As proposed
Lighting, interior	The interior lighting power shall be determined in accordance with Section C405.3.2. Where the occupancy of the building is not known, the lighting power density shall be 1.0 Watt per square foot (10.7 W/m^2) based on the categorization of buildings with unknown space classification as offices.	As proposed
Lighting, exterior	The lighting power shall be determined in accordance with Table C405.4.2(2) and C405.4.2(3). Areas and dimensions of surfaces shall be the same as proposed.	As proposed

(continued)

TABLE C407.5.1(1)—continued
SPECIFICATIONS FOR THE STANDARD REFERENCE AND PROPOSED DESIGNS

BUILDING COMPONENT CHARACTERISTICS	STANDARD REFERENCE DESIGN	PROPOSED DESIGN
Internal gains	Same as proposed	Receptacle, motor and process loads shall be modeled and estimated based on the space use classification. End-use load components within and associated with the building shall be modeled to include, but not be limited to, the following: exhaust fans, parking garage ventilation fans, exterior building lighting, swimming pool heaters and pumps, elevators, escalators, refrigeration equipment and cooking equipment.
Schedules	Same as proposed **Exception:** Thermostat settings and schedules for HVAC systems that utilize radiant heating, radiant cooling and elevated air speed, provided that equivalent levels of occupant thermal comfort are demonstrated by means of equal Standard Effective Temperature as calculated in Normative Appendix B of ASHRAE Standard 55.	Operating schedules shall include hourly profiles for daily operation and shall account for variations between weekdays, weekends, holidays and any seasonal operation. Schedules shall model the time-dependent variations in occupancy, illumination, receptacle loads, thermostat settings, mechanical ventilation, HVAC equipment availability, service hot water usage and any process loads. The schedules shall be typical of the proposed building type as determined by the designer and approved by the jurisdiction.
Mechanical ventilation	Same as proposed	As proposed, in accordance with Section C403.2.2.
Heating systems	Fuel type: same as proposed design	As proposed
	Equipment type[a]: as specified in Tables C407.5.1(2) and C407.5.1(3)	As proposed
	Efficiency: as specified in Tables C403.3.2(4) and C403.3.2(5)	As proposed
	Capacity[b]: sized proportionally to the capacities in the proposed design based on sizing runs, and shall be established such that no smaller number of unmet heating load hours and no larger heating capacity safety factors are provided than in the proposed design.	As proposed
Cooling systems	Fuel type: same as proposed design	As proposed
	Equipment type[c]: as specified in Tables C407.5.1(2) and C407.5.1(3)	As proposed
	Efficiency: as specified in Tables C403.3.2(1), C403.3.2(2) and C403.3.2(3)	As proposed
	Capacity[b]: sized proportionally to the capacities in the proposed design based on sizing runs, and shall be established such that no smaller number of unmet cooling load hours and no larger cooling capacity safety factors are provided than in the proposed design.	As proposed
	Economizer[d]: same as proposed, in accordance with Section C403.5.	As proposed
Service water heating[e]	Fuel type: same as proposed	As proposed
	Efficiency: as specified in Table C404.2	For *Group R*, as proposed multiplied by SWHF. For other than *Group R*, as proposed multiplied by efficiency as provided by the manufacturer of the DWHR unit.
	Capacity: same as proposed Where no service water hot water system exists or is specified in the proposed design, no service hot water heating shall be modeled.	As proposed

SWHF = Service water heat recovery factor, DWHR = Drain water heat recovery.

a. Where no heating system exists or has been specified, the heating system shall be modeled as fossil fuel. The system characteristics shall be identical in both the standard reference design and proposed design.

b. The ratio between the capacities used in the annual simulations and the capacities determined by sizing runs shall be the same for both the standard reference design and proposed design.

c. Where no cooling system exists or no cooling system has been specified, the cooling system shall be modeled as an air-cooled single-zone system, one unit per thermal zone. The system characteristics shall be identical in both the standard reference design and proposed design.

d. If an economizer is required in accordance with Table C403.5(1) and where no economizer exists or is specified in the proposed design, then a supply-air economizer shall be provided in the standard reference design in accordance with Section C403.5.

e. The SWHF shall be applied as follows:

1. Where potable water from the DWHR unit supplies not less than one shower and not greater than two showers, of which the drain water from the same showers flows through the DWHR unit then SWHF = [1 – (DWHR unit efficiency • 0.36)].
2. Where potable water from the DWHR unit supplies not less than three showers and not greater than four showers, of which the drain water from the same showers flows through the DWHR unit then SWHF = [1 – (DWHR unit efficiency • 0.33)].
3. Where potable water from the DWHR unit supplies not less than five showers and not greater than six showers, of which the drain water from the same showers flows through the DWHR unit, then SWHF = [1 – (DWHR unit efficiency • 0.26)].
4. Where Items 1 through 3 are not met, SWHF = 1.0.

TABLE C407.5.1(2)
HVAC SYSTEMS MAP

CONDENSER COOLING SOURCE[a]	HEATING SYSTEM CLASSIFICATION[b]	STANDARD REFERENCE DESIGN HVC SYSTEM TYPE[c]		
		Single-zone Residential System	Single-zone Nonresidential System	All Other
Water/ground	Electric resistance	System 5	System 5	System 1
	Heat pump	System 6	System 6	System 6
	Fossil fuel	System 7	System 7	System 2
Air/none	Electric resistance	System 8	System 9	System 3
	Heat pump	System 8	System 9	System 3
	Fossil fuel	System 10	System 11	System 4

a. Select "water/ground" where the proposed design system condenser is water or evaporatively cooled; select "air/none" where the condenser is air cooled. Closed-circuit dry coolers shall be considered to be air cooled. Systems utilizing district cooling shall be treated as if the condenser water type were "water." Where mechanical cooling is not specified or the mechanical cooling system in the proposed design does not require heat rejection, the system shall be treated as if the condenser water type were "Air." For proposed designs with ground-source or groundwater-source heat pumps, the standard reference design HVAC system shall be water-source heat pump (System 6).

b. Select the path that corresponds to the proposed design heat source: electric resistance, heat pump (including air source and water source), or fuel fired. Systems utilizing district heating (steam or hot water) and systems without heating capability shall be treated as if the heating system type were "fossil fuel." For systems with mixed fuel heating sources, the system or systems that use the secondary heating source type (the one with the smallest total installed output capacity for the spaces served by the system) shall be modeled identically in the standard reference design and the primary heating source type shall be used to determine standard reference design HVAC system type.

c. Select the standard reference design HVAC system category: The system under "single-zone residential system" shall be selected where the HVAC system in the proposed design is a single-zone system and serves a Group R occupancy. The system under "single-zone nonresidential system" shall be selected where the HVAC system in the proposed design is a single-zone system and serves other than Group R occupancy. The system under "all other" shall be selected for all other cases.

TABLE C407.5.1(3)
SPECIFICATIONS FOR THE STANDARD REFERENCE DESIGN HVAC SYSTEM DESCRIPTIONS

SYSTEM NO.	SYSTEM TYPE	FAN CONTROL	COOLING TYPE	HEATING TYPE
1	Variable air volume with parallel fan-powered boxes[a]	VAV[d]	Chilled water[e]	Electric resistance
2	Variable air volume with reheat[b]	VAV[d]	Chilled water[e]	Hot water fossil fuel boiler[f]
3	Packaged variable air volume with parallel fan-powered boxes[a]	VAV[d]	Direct expansion[c]	Electric resistance
4	Packaged variable air volume with reheat[b]	VAV[d]	Direct expansion[c]	Hot water fossil fuel boiler[f]
5	Two-pipe fan coil	Constant volume[i]	Chilled water[e]	Electric resistance
6	Water-source heat pump	Constant volume[i]	Direct expansion[c]	Electric heat pump and boiler[g]
7	Four-pipe fan coil	Constant volume[i]	Chilled water[e]	Hot water fossil fuel boiler[f]
8	Packaged terminal heat pump	Constant volume[i]	Direct expansion[c]	Electric heat pump[h]
9	Packaged rooftop heat pump	Constant volume[i]	Direct expansion[c]	Electric heat pump[h]
10	Packaged terminal air conditioner	Constant volume[i]	Direct expansion	Hot water fossil fuel boiler[f]
11	Packaged rooftop air conditioner	Constant volume[i]	Direct expansion	Fossil fuel furnace

For SI: 1 foot = 304.8 mm, 1 cfm/ft^2 = 0.4719 L/s, 1 Btu/h = 0.293/W, °C = [(°F) - 32]/1.8.

a. **VAV with parallel boxes:** Fans in parallel VAV fan-powered boxes shall be sized for 50 percent of the peak design flow rate and shall be modeled with 0.35 W/cfm fan power. Minimum volume setpoints for fan-powered boxes shall be equal to the minimum rate for the space required for ventilation consistent with Section C403.6.1, Item 3. Supply air temperature setpoint shall be constant at the design condition.

b. **VAV with reheat:** Minimum volume setpoints for VAV reheat boxes shall be 0.4 cfm/ft^2 of floor area. Supply air temperature shall be reset based on zone demand from the design temperature difference to a 10°F temperature difference under minimum load conditions. Design airflow rates shall be sized for the reset supply air temperature, i.e., a 10°F temperature difference.

c. **Direct expansion:** The fuel type for the cooling system shall match that of the cooling system in the proposed design.

d. **VAV:** Where the proposed design system has a supply, return or relief fan motor 25 hp or larger, the corresponding fan in the VAV system of the standard reference design shall be modeled assuming a variable-speed drive. For smaller fans, a forward-curved centrifugal fan with inlet vanes shall be modeled. Where the proposed design's system has a direct digital control system at the zone level, static pressure setpoint reset based on zone requirements in accordance with Section C403.8.5 shall be modeled.

e. **Chilled water:** For systems using purchased chilled water, the chillers are not explicitly modeled and chilled water costs shall be based as determined in Sections C407.3 and C407.5.2. Otherwise, the standard reference design's chiller plant shall be modeled with chillers having the number as indicated in Table C407.5.1(4) as a function of standard reference building chiller plant load and type as indicated in Table C407.5.1(5) as a function of individual chiller load. Where chiller fuel source is mixed, the system in the standard reference design shall have chillers with the same fuel types and with capacities having the same proportional capacity as the proposed design's chillers for each fuel type. Chilled water supply temperature shall be modeled at 44°F design supply temperature and 56°F return temperature. Piping losses shall not be modeled in either building model. Chilled water supply water temperature shall be reset in accordance with Section C403.9.3. Pump system power for each pumping system shall be the same as the proposed design; where the proposed design has no chilled water pumps, the standard reference design pump power shall be 22 W/gpm (equal to a pump operating against a 75-foot head, 65-percent combined impeller and motor efficiency). The chilled water system shall be modeled as primary-only variable flow with flow maintained at the design rate through each chiller using a bypass. Chilled water pumps shall be modeled as riding the pump curve or with variable-speed drives where required in Section C403.9.3. The heat rejection device shall be an axial fan cooling tower with two-speed fans where required in Section C403.9. Condenser water design supply temperature shall be 85°F or 10°F approach to design wet-bulb temperature, whichever is lower, with a design temperature rise of 10°F. The tower shall be controlled to maintain a 70°F leaving water temperature where weather permits, floating up to leaving water temperature at design conditions. Pump system power for each pumping system shall be the same as the proposed design; where the proposed design has no condenser water pumps, the standard reference design pump power shall be 19 W/gpm (equal to a pump operating against a 60-foot head, 60-percent combined impeller and motor efficiency). Each chiller shall be modeled with separate condenser water and chilled water pumps interlocked to operate with the associated chiller.

f. **Fossil fuel boiler:** For systems using purchased hot water or steam, the boilers are not explicitly modeled and hot water or steam costs shall be based on actual utility rates. Otherwise, the boiler plant shall use the same fuel as the proposed design and shall be natural draft. The standard reference design boiler plant shall be modeled with a single boiler where the standard reference design plant load is 600,000 Btu/h and less and with two equally sized boilers for plant capacities exceeding 600,000 Btu/h. Boilers shall be staged as required by the load. Hot water supply temperature shall be modeled at 180°F design supply temperature and 130°F return temperature. Piping losses shall not be modeled in either building model. Hot water supply water temperature shall be reset in accordance with Section C403.9.3. Pump system power for each pumping system shall be the same as the proposed design; where the proposed design has no hot water pumps, the standard reference design pump power shall be 19 W/gpm (equal to a pump operating against a 60-foot head, 60-percent combined impeller and motor efficiency). The hot water system shall be modeled as primary only with continuous variable flow. Hot water pumps shall be modeled as riding the pump curve or with variable speed drives where required by Section C403.9.3.

g. **Electric heat pump and boiler:** Water-source heat pumps shall be connected to a common heat pump water loop controlled to maintain temperatures between 60°F and 90°F. Heat rejection from the loop shall be provided by an axial fan closed-circuit evaporative fluid cooler with two-speed fans where required in Section C403.8.5. Heat addition to the loop shall be provided by a boiler that uses the same fuel as the proposed design and shall be natural draft. Where no boilers exist in the proposed design, the standard reference building boilers shall be fossil fuel. The standard reference design boiler plant shall be modeled with a single boiler where the standard reference design plant load is 600,000 Btu/h or less and with two equally sized boilers for plant capacities exceeding 600,000 Btu/h. Boilers shall be staged as required by the load. Piping losses shall not be modeled in either building model. Pump system power shall be the same as the proposed design; where the proposed design has no pumps, the standard reference design pump power shall be 22 W/gpm, which is equal to a pump operating against a 75-foot head, with a 65-percent combined impeller and motor efficiency. Loop flow shall be variable with flow shutoff at each heat pump when its compressor cycles off as required by Section C403.9.3. Loop pumps shall be modeled as riding the pump curve or with variable speed drives where required by Section C403.9.3.

h. **Electric heat pump:** Electric air-source heat pumps shall be modeled with electric auxiliary heat. The system shall be controlled with a multistage space thermostat and an outdoor air thermostat wired to energize auxiliary heat only on the last thermostat stage and when outdoor air temperature is less than 40°F.

i. **Constant volume:** Fans shall be controlled in the same manner as in the proposed design; i.e., fan operation whenever the space is occupied or fan operation cycled on calls for heating and cooling. Where the fan is modeled as cycling and the fan energy is included in the energy efficiency rating of the equipment, fan energy shall not be modeled explicitly.

TABLE C407.5.1(4)
NUMBER OF CHILLERS

TOTAL CHILLER PLANT CAPACITY	NUMBER OF CHILLERS
≤ 300 tons	1
> 300 tons, < 600 tons	2, sized equally
≥ 600 tons	2 minimum, with chillers added so that all are sized equally and none is larger than 800 tons

For SI: 1 ton = 3517 W.

TABLE C407.5.1(5)
WATER CHILLER TYPES

INDIVIDUAL CHILLER PLANT CAPACITY	ELECTRIC CHILLER TYPE	FOSSIL FUEL CHILLER TYPE
≤ 100 tons	Reciprocating	Single-effect absorption, direct fired
> 100 tons, < 300 tons	Screw	Double-effect absorption, direct fired
≥ 300 tons	Centrifugal	Double-effect absorption, direct fired

For SI: 1 ton = 3517 W.

SECTION C408 MAINTENANCE INFORMATION AND SYSTEM COMMISSIONING

C408.1 General. This section covers the provision of maintenance information and the commissioning of, and the functional testing requirements for, building systems.

C408.1.1 Building operations and maintenance information. The building operations and maintenance documents shall be provided to the owner and shall consist of manufacturers' information, specifications and recommendations; programming procedures and data points; narratives; and other means of illustrating to the owner how the building, equipment and systems are intended to be installed, maintained and operated. Required regular maintenance actions for equipment and systems shall be clearly stated on a readily visible label. The label shall include the title or publication number for the operation and maintenance manual for that particular model and type of product.

C408.2 Mechanical systems and service water-heating systems commissioning and completion requirements. Prior to the final mechanical and plumbing inspections, the *registered design professional or approved agency* shall provide evidence of mechanical systems *commissioning* and completion in accordance with the provisions of this section.

Construction document notes shall clearly indicate provisions for *commissioning* and completion requirements in accordance with this section and are permitted to refer to specifications for further requirements. Copies of all documentation shall be given to the owner or owner's authorized agent and made available to the *code official* upon request in accordance with Sections C408.2.4 and C408.2.5.

Exceptions: The following systems are exempt:

1. Mechanical systems and service water heater systems in buildings where the total mechanical equipment capacity is less than 480,000 Btu/h (140.7 kW) cooling capacity and 600,000 Btu/h (175.8 kW) combined service water-heating and space-heating capacity.
2. Systems included in Section C403.5 that serve individual *dwelling units* and *sleeping units*.

C408.2.1 Commissioning plan. A *commissioning plan* shall be developed by a *registered design professional* or *approved agency* and shall include the following items:

1. A narrative description of the activities that will be accomplished during each phase of *commissioning*, including the personnel intended to accomplish each of the activities.
2. A listing of the specific equipment, appliances or systems to be tested and a description of the tests to be performed.
3. Functions to be tested including, but not limited to, calibrations and economizer controls.
4. Conditions under which the test will be performed. Testing shall affirm winter and summer design conditions and full outside air conditions.
5. Measurable criteria for performance.

C408.2.2 Systems adjusting and balancing. HVAC systems shall be balanced in accordance with generally accepted engineering standards. Air and water flow rates shall be measured and adjusted to deliver final flow rates within the tolerances provided in the product specifications. Test and balance activities shall include air system and hydronic system balancing.

C408.2.2.1 Air systems balancing. Each supply air outlet and *zone* terminal device shall be equipped with means for air balancing in accordance with the requirements of Chapter 6 of the *International Mechanical Code*. Discharge dampers used for air-system balancing are prohibited on constant-volume fans and variable-volume fans with motors 10 hp (18.6 kW) and larger.

Air systems shall be balanced in a manner to first minimize throttling losses then, for fans with system power of greater than 1 hp (0.746 kW), fan speed shall be adjusted to meet design flow conditions.

Exception: Fans with fan motors of 1 hp (0.74 kW) or less are not required to be provided with a means for air balancing.

C408.2.2.2 Hydronic systems balancing. Individual hydronic heating and cooling coils shall be equipped with means for balancing and measuring flow. Hydronic systems shall be proportionately balanced in a manner to first minimize throttling losses, then the pump impeller shall be trimmed or pump speed shall be adjusted to meet design flow conditions. Each hydronic system shall have either the capability to measure pressure across the pump, or test ports at each side of each pump.

Exception: The following equipment is not required to be equipped with a means for balancing or measuring flow:

1. Pumps with pump motors of 5 hp (3.7 kW) or less.
2. Where throttling results in not greater than 5 percent of the nameplate horsepower draw above that required if the impeller were trimmed.

C408.2.3 Functional performance testing. Functional performance testing specified in Sections C408.2.3.1 through C408.2.3.3 shall be conducted.

C408.2.3.1 Equipment. Equipment functional performance testing shall demonstrate the installation and operation of components, systems, and system-to-system interfacing relationships in accordance with approved plans and specifications such that operation, function, and maintenance serviceability for each of the commissioned systems is confirmed. Testing shall include all modes and *sequence of operation*, including under full-load, part-load and the following emergency conditions:

1. All modes as described in the *sequence* of *operation*.
2. Redundant or *automatic* back-up mode.
3. Performance of alarms.
4. Mode of operation upon a loss of power and restoration of power.

Exception: Unitary or packaged HVAC equipment listed in Tables C403.3.2(1) through C403.3.2(3) that do not require supply air economizers.

C408.2.3.2 Controls. HVAC and service water-heating control systems shall be tested to document that control devices, components, equipment and systems are calibrated and adjusted and operate in accordance with approved plans and specifications. Sequences of operation shall be functionally tested to document they operate in accordance with *approved* plans and specifications.

C408.2.3.3 Economizers. Air economizers shall undergo a functional test to determine that they operate in accordance with manufacturer's specifications.

C408.2.4 Preliminary commissioning report. A preliminary report of *commissioning* test procedures and results shall be completed and certified by the *registered design professional* or *approved agency* and provided to the building owner or owner's authorized agent. The report shall be organized with mechanical and service hot water findings in separate sections to allow independent review. The report shall be identified as "Preliminary Commissioning Report," shall include the completed Commissioning Compliance Checklist, Figure C408.2.4, and shall identify:

1. Itemization of deficiencies found during testing required by this section that have not been corrected at the time of report preparation.
2. Deferred tests that cannot be performed at the time of report preparation because of climatic conditions.
3. Climatic conditions required for performance of the deferred tests.
4. Results of functional performance tests.
5. Functional performance test procedures used during the commissioning process, including measurable criteria for test acceptance.

C408.2.4.1 Acceptance of report. Buildings, or portions thereof, shall not be considered as acceptable for a final inspection pursuant to Section C105.2.6 until the *code official* has received the Preliminary Commissioning Report from the building owner or owner's authorized agent.

C408.2.4.2 Copy of report. The *code official* shall be permitted to require that a copy of the Preliminary Commissioning Report be made available for review by the *code official.*

C408.2.5 Documentation requirements. The *construction documents* shall specify that the documents described in this section be provided to the building owner or owner's authorized agent within 90 days of the date of receipt of the *certificate of occupancy*.

C408.2.5.1 System balancing report. A written report describing the activities and measurements completed in accordance with Section C408.2.2.

C408.2.5.2 Final commissioning report. A report of test procedures and results identified as "Final Commissioning Report" shall be delivered to the building owner or owner's authorized agent. The report shall be organized with mechanical system and service hot water system findings in separate sections to allow independent review. The report shall include the following:

1. Results of functional performance tests.
2. Disposition of deficiencies found during testing, including details of corrective measures used or proposed.

3. Functional performance test procedures used during the commissioning process including measurable criteria for test acceptance, provided herein for repeatability.

Exception: Deferred tests that cannot be performed at the time of report preparation due to climatic conditions.

C408.3 Functional testing of lighting controls. Automatic lighting controls required by this code shall comply with this section.

C408.3.1 Functional testing. Prior to passing final inspection, the *registered design professional* shall provide evidence that the lighting control systems have been tested to ensure that control hardware and software are calibrated, adjusted, programmed and in proper working condition in accordance with the *construction documents* and manufacturer's instructions. Functional testing shall be in accordance with Sections C408.3.1.1 through C408.3.1.3 for the applicable control type.

C408.3.1.1 Occupant sensor controls. Where *occupant sensor controls* are provided, the following procedures shall be performed:

1. Certify that the *occupant sensor* has been located and aimed in accordance with manufacturer recommendations.
2. For projects with seven or fewer *occupant sensors*, each sensor shall be tested.
3. For projects with more than seven *occupant sensors,* testing shall be done for each unique combination of sensor type and space geometry. Where multiples of each unique combination of sensor type and space geometry are provided, not less than 10 percent and in no case fewer than one, of each combination shall be tested unless the *code official* or design professional requires a higher percentage to be tested. Where 30 percent or more of the tested controls fail, all remaining identical combinations shall be tested.

   For *occupant sensor controls* to be tested, verify the following:

   3.1. Where *occupant sensor controls* include status indicators, verify correct operation.

   3.2. The controlled lights turn off or down to the permitted level within the required time.

   3.3. For auto-on *occupant sensor controls*, the lights turn on to the permitted level when an occupant enters the space.

   3.4. For manual-on *occupant sensor controls*, the lights turn on only when manually activated.

   3.5. The lights are not incorrectly turned on by movement in adjacent areas or by HVAC operation.

C408.3.1.2 Time-switch controls. Where *time-switch controls* are provided, the following procedures shall be performed:

1. Confirm that the *time-switch control* is programmed with accurate weekday, weekend and holiday schedules.
2. Provide documentation to the owner of *time-switch controls* programming including weekday, weekend, holiday schedules, and set-up and preference program settings.
3. Verify the correct time and date in the time switch.
4. Verify that any battery back-up is installed and energized.
5. Verify that the override time limit is set to not more than 2 hours.
6. Simulate occupied condition. Verify and document the following:

   6.1. All lights can be turned on and off by their respective area control switch.

   6.2. The switch only operates lighting in the enclosed space in which the switch is located.

7. Simulate unoccupied condition. Verify and document the following:

   7.1. Nonexempt lighting turns off.

   7.2. Manual override switch allows only the lights in the enclosed space where the override switch is located to turn on or remain on until the next scheduled shutoff occurs.

8. Additional testing as specified by the *registered design professional.*

C408.3.1.3 Daylight responsive controls. Where *daylight responsive controls* are provided, the following shall be verified:

1. Control devices have been properly located, field calibrated and set for accurate setpoints and threshold light levels.
2. Daylight controlled lighting loads adjust to light level setpoints in response to available daylight.
3. The calibration adjustment equipment is located for *ready access* only by authorized personnel.

C408.3.2 Documentation requirements. The *construction documents* shall specify that the documents described in this section be provided to the building owner or owner's authorized agent within 90 days of the date of receipt of the *certificate of occupancy*.

C408.3.2.1 Drawings. Construction documents shall include the location and catalogue number of each piece of equipment.

C408.3.2.2 Manuals. An operating and maintenance manual shall be provided and include the following:

1. Name and address of not less than one service agency for installed equipment.

2. A narrative of how each system is intended to operate, including recommended setpoints.
3. Submittal data indicating all selected options for each piece of lighting equipment and lighting controls.
4. Operation and maintenance manuals for each piece of lighting equipment. Required routine maintenance actions, cleaning and recommended relamping shall be clearly identified.
5. A schedule for inspecting and recalibrating all lighting controls.

C408.3.2.3 Report. A report of test results shall be provided and include the following:

1. Results of functional performance tests.
2. Disposition of deficiencies found during testing, including details of corrective measures used or proposed.

Project Information: ______________________________ Project Name: ____________________________________

Project Address: ____

Commissioning Authority: ____

Commissioning Plan (Section C408.2.1)

☐ Commissioning Plan was used during construction and includes all items required by Section C408.2.1

☐ Systems Adjusting and Balancing has been completed.

☐ HVAC Equipment Functional Testing has been executed. If applicable, deferred and follow-up testing is scheduled to be provided on: ____________________________

☐ HVAC Controls Functional Testing has been executed. If applicable, deferred and follow-up testing is scheduled to be provided on: ____________________________

☐ Economizer Functional Testing has been executed. If applicable, deferred and follow-up testing is scheduled to be provided on: ____________________________

☐ Lighting Controls Functional Testing has been executed. If applicable, deferred and follow-up testing is scheduled to be provided on: ____________________________

☐ Service Water Heating System Functional Testing has been executed. If applicable, deferred and follow-up testing is scheduled to be provided on: ____________________________

☐ Manual, record documents and training have been completed or scheduled

☐ Preliminary Commissioning Report submitted to owner and includes all items required by Section C408.2.4

I hereby certify that the commissioning provider has provided me with evidence of mechanical, service water heating and lighting systems commissioning in accordance with the 2018 IECC.

Signature of Building Owner or Owner's Representative ____________________________ Date______________

FIGURE C408.2.4
COMMISSIONING COMPLIANCE CHECKLIST

CHAPTER 5 [CE]

EXISTING BUILDINGS

User note:

About this chapter: *Many buildings are renovated or altered in numerous ways that could affect the energy use of the building as a whole. Chapter 5 requires the application of certain parts of Chapter 4 in order to maintain, if not improve, the conservation of energy by the renovated or altered building.*

SECTION C501
GENERAL

C501.1 Scope. The provisions of this chapter shall control the *alteration, repair, addition* and *change of occupancy* of existing buildings and structures.

C501.2 Existing buildings. Except as specified in this chapter, this code shall not be used to require the removal, *alteration* or abandonment of, nor prevent the continued use and maintenance of, an existing *building* or *building* system lawfully in existence at the time of adoption of this code.

C501.3 Maintenance. *Buildings* and structures, and parts thereof, shall be maintained in a safe and sanitary condition. Devices and systems required by this code shall be maintained in conformance to the code edition under which they were installed. The owner or the owner's authorized agent shall be responsible for the maintenance of buildings and structures. The requirements of this chapter shall not provide the basis for removal or abrogation of energy conservation, fire protection and safety systems and devices in existing structures.

C501.4 Compliance. *Alterations, repairs, additions* and changes of occupancy to, or relocation of, existing *buildings* and structures shall comply with the provisions for *alterations, repairs, additions* and changes of occupancy or relocation, respectively, in this code and in the *International Building Code, International Existing Building Code, International Fire Code, International Fuel Gas Code, International Mechanical Code, International Plumbing Code, International Property Maintenance Code, International Private Sewage Disposal Code* and NFPA 70.

C501.5 New and replacement materials. Except as otherwise required or permitted by this code, materials permitted by the applicable code for new construction shall be used. Like materials shall be permitted for *repairs*, provided that hazards to life, health or property are not created. Hazardous materials shall not be used where the code for new construction would not allow use of these materials in buildings of similar occupancy, purpose and location.

C501.6 Historic buildings. Provisions of this code relating to the construction, *repair, alteration*, restoration and movement of structures, and *change of occupancy* shall not be mandatory for *historic buildings* provided that a report has been submitted to the *code official* and signed by a *registered design professional*, or a representative of the State Historic Preservation Office or the historic preservation authority having jurisdiction, demonstrating that compliance with that provision would threaten, degrade or destroy the historic form, fabric or function of the *building*.

SECTION C502
ADDITIONS

C502.1 General. *Additions* to an existing *building, building* system or portion thereof shall conform to the provisions of this code as those provisions relate to new construction without requiring the unaltered portion of the existing *building* or *building* system to comply with this code. *Additions* shall not create an unsafe or hazardous condition or overload existing *building* systems. An *addition* shall be deemed to comply with this code if the *addition* alone complies or if the existing building and *addition* comply with this code as a single building. *Additions* shall comply with Sections C402, C403, C404, C405 and C502.2.

Additions complying with ANSI/ASHRAE/IESNA 90.1. need not comply with Sections C402, C403, C404 and C405.

C502.2 Prescriptive compliance. *Additions* shall comply with Sections C502.2.1 through C502.2.6.2.

C502.2.1 Vertical fenestration. New *vertical fenestration* area that results in a total building *fenestration* area less than or equal to that specified in Section C402.4.1 shall comply with Section C402.1.5, C402.4.3 or C407. *Additions* with *vertical fenestration* that result in a total building *fenestration* area greater than Section C402.4.1 or *additions* that exceed the fenestration area greater than Section C402.4.1 shall comply with Section C402.4.1.1 for the *addition* only. *Additions* that result in a total building vertical fenestration area exceeding that specified in Section C402.4.1.1 shall comply with Section C402.1.5 or C407

C502.2.2 Skylight area. New *skylight* area that results in a total building *fenestration* area less than or equal to that specified in Section C402.4.1 shall comply with Section C402.1.5 or C407. *Additions* with *skylight* area that result in a total building *skylight* area greater than C402.4.1 or additions that exceed the *skylight* area shall comply with Section C402.4.1.2 for the *addition* only. *Additions* that result in a total building *skylight* area exceeding that specified in Section C402.4.1.2 shall comply with Section C402.1.5 or C407.

C502.2.3 Building mechanical systems. New mechanical systems and equipment that are part of the *addition* and serve the building heating, cooling and ventilation needs shall comply with Section C403.

C502.2.4 Service water-heating systems. New service water-heating equipment, controls and service water heating piping shall comply with Section C404.

C502.2.5 Pools and inground permanently installed spas. New pools and inground permanently installed spas shall comply with Section C404.10.

C502.2.6 Lighting power and systems. New lighting systems that are installed as part of the addition shall comply with Section C405.

C502.2.6.1 Interior lighting power. The total interior lighting power for the *addition* shall comply with Section C405.3.2 for the *addition* alone, or the existing building and the *addition* shall comply as a single building.

C502.2.6.2 Exterior lighting power. The total exterior lighting power for the *addition* shall comply with Section C405.4.2 for the *addition* alone, or the existing building and the *addition* shall comply as a single building.

SECTION C503
ALTERATIONS

C503.1 General. *Alterations* to any *building* or structure shall comply with the requirements of Section C503 and the code for new construction. *Alterations* shall be such that the existing *building* or structure is not less conforming to the provisions of this code than the existing *building* or structure was prior to the *alteration*. *Alterations* to an existing *building*, *building* system or portion thereof shall conform to the provisions of this code as those provisions relate to new construction without requiring the unaltered portions of the existing *building* or *building* system to comply with this code. *Alterations* shall not create an unsafe or hazardous condition or overload existing *building* systems.

Alterations complying with ANSI/ASHRAE/IESNA 90.1. need not comply with Sections C402, C403, C404 and C405.

Exception: The following *alterations* need not comply with the requirements for new construction, provided that the energy use of the building is not increased:

1. Storm windows installed over existing *fenestration*.
2. Surface-applied window film installed on existing single-pane *fenestration* assemblies reducing solar heat gain, provided that the code does not require the glazing or *fenestration* to be replaced.
3. Existing ceiling, wall or floor cavities exposed during construction, provided that these cavities are filled with insulation.
4. Construction where the existing roof, wall or floor cavity is not exposed.
5. *Roof recover*.
6. *Air barriers* shall not be required for *roof recover* and roof replacement where the *alterations* or renovations to the building do not include *alterations*, renovations or *repairs* to the remainder of the building envelope.

C503.2 Change in space conditioning. Any nonconditioned or low-energy space that is altered to become *conditioned space* shall be required to be brought into full compliance with this code.

Exceptions:

1. Where the component performance alternative in Section C402.1.5 is used to comply with this section, the proposed UA shall be not greater than 110 percent of the target UA.
2. Where the total building performance option in Section C407 is used to comply with this section, the annual energy cost of the proposed design shall be not greater than 110 percent of the annual energy cost otherwise permitted by Section C407.3.

C503.3 Building envelope. New building envelope assemblies that are part of the *alteration* shall comply with Sections C402.1 through C402.5.

Exception: Where the existing building exceeds the fenestration area limitations of Section C402.4.1 prior to alteration, the building is exempt from Section C402.4.1 provided that there is not an increase in fenestration area.

C503.3.1 Roof replacement. *Roof replacements* shall comply with Section C402.1.3, C402.1.4, C402.1.5 or C407 where the existing roof assembly is part of the *building thermal envelope* and contains insulation entirely above the roof deck.

C503.3.2 Vertical fenestration. The addition of *vertical fenestration* that results in a total building *fenestration* area less than or equal to that specified in Section C402.4.1 shall comply with Section C402.1.5, C402.4.3 or C407. The addition of *vertical fenestration* that results in a total building *fenestration* area greater than Section C402.4.1 shall comply with Section C402.4.1.1 for the space adjacent to the new fenestration only. *Alterations* that result in a total building *vertical fenestration* area exceeding that specified in Section C402.4.1.1 shall comply with Section C402.1.5 or C407. Provided that the vertical fenestration area is not changed, using the same vertical fenestration area in the *standard reference design* as the building prior to alteration shall be an alternative to using the vertical fenestration area specified in Table C407.5.1(1).

C503.3.3 Skylight area. New *skylight* area that results in a total building *skylight* area less than or equal to that specified in Section C402.4.1 shall comply with Section C402.1.5, C402.4 or C407. The addition of *skylight* area that results in a total building skylight area greater than Section C402.4.1 shall comply with Section C402.4.1.2 for the space adjacent to the new skylights. *Alterations* that result in a total building skylight area exceeding that specified in Section C402.4.1.2 shall comply with Section C402.1.5 or C407. Provided that the skylight area is not changed, using the same skylight area in the *standard reference design* as the building prior to alteration shall be an alternative to using the skylight area specified in Table C407.5.1(1).

C503.4 Heating and cooling systems. New heating, cooling and duct systems that are part of the *alteration* shall comply with Sections C403.

C503.4.1 Economizers. New cooling systems that are part of *alteration* shall comply with Section C403.5.

C503.5 Service hot water systems. New service hot water systems that are part of the *alteration* shall comply with Section C404.

C503.6 Lighting systems. New lighting systems that are part of the *alteration* shall comply with Section C405.

Exception. *Alterations* that replace less than 10 percent of the luminaires in a space, provided that such *alterations* do not increase the installed interior lighting power.

SECTION C504
REPAIRS

C504.1 General. *Buildings* and structures, and parts thereof, shall be repaired in compliance with Section C501.3 and this section. Work on nondamaged components that is necessary for the required *repair* of damaged components shall be considered to be part of the *repair* and shall not be subject to the requirements for *alterations* in this chapter. Routine maintenance required by Section C501.3, ordinary *repairs* exempt from *permit* and abatement of wear due to normal service conditions shall not be subject to the requirements for *repairs* in this section.

Where a building was constructed to comply with ANSI/ASHRAE/IESNA 90.1, repairs shall comply with the standard and need not comply with Sections C402, C403, C404 and C405.

C504.2 Application. For the purposes of this code, the following shall be considered to be repairs:

1. Glass-only replacements in an existing sash and frame.
2. *Roof repairs.*
3. Air barriers shall not be required for *roof repair* where the repairs to the building do not include *alterations*, renovations or *repairs* to the remainder of the building envelope.
4. Replacement of existing doors that separate conditioned space from the exterior shall not require the installation of a vestibule or revolving door, provided that an existing vestibule that separates a conditioned space from the exterior shall not be removed.
5. *Repairs* where only the bulb, the ballast or both within the existing luminaires in a space are replaced, provided that the replacement does not increase the installed interior lighting power.

SECTION C505
CHANGE OF OCCUPANCY OR USE

C505.1 General. Spaces undergoing a change in occupancy that would result in an increase in demand for either fossil fuel or electrical energy shall comply with this code. Where the use in a space changes from one use in Table C405.3.2(1) or C405.3.2(2) to another use in Table C405.3.2(1) or C405.3.2(2), the installed lighting wattage shall comply with Section C405.3. Where the space undergoing a change in occupancy or use is in a building with a fenestration area that exceeds the limitations of Section C402.4.1, the space is exempt from Section C402.4.1 provided that there is not an increase in fenestration area.

Exceptions:

1. Where the component performance alternative in Section C402.1.5 is used to comply with this section, the proposed UA shall be not greater than 110 percent of the target UA.
2. Where the total building performance option in Section C407 is used to comply with this section, the annual energy cost of the proposed design shall be not greater than 110 percent of the annual energy cost otherwise permitted by Section C407.3.

CHAPTER 6 [CE]

REFERENCED STANDARDS

User note:

About this chapter: *Chapter 6 lists the full title, edition year and address of the promulgator for all standards that are referenced in the code. The section numbers in which the standards are referenced are also listed.*

This chapter lists the standards that are referenced in various sections of this document. The standards are listed herein by the promulgating agency of the standard, the standard identification, the effective date and title, and the section or sections of this document that reference the standard. The application of the referenced standards shall be as specified in Section 107.

AAMA

American Architectural Manufacturers Association
1827 Walden Office Square
Suite 550
Schaumburg, IL 60173-4268

AAMA/WDMA/CSA 101/I.S.2/A C440—17: North American Fenestration Standard/Specifications for Windows, Doors and Unit Skylights
Table C402.5.2

AHAM

Association of Home Appliance Manufacturers
1111 19th Street NW, Suite 402
Washington, DC 20036

ANSI/AHAM RAC-1—2008: Room Air Conditioners
Table C403.3.2(3)

AHAM HRF-1—2016: Energy, Performance and Capacity of Household Refrigerators, Refrigerator-Freezers and Freezers
Table C403.10.1

AHRI

Air-Conditioning, Heating, & Refrigeration Institute
2111 Wilson Blvd, Suite 500
Arlington, VA 22201

ISO/AHRI/ASHRAE 13256-1 (1998 RA2014): Water-to-Air and Brine-to-Air Heat Pumps—Testing and Rating for Performance
Table C403.3.2(2)

ISO/AHRI/ASHRAE 13256-2 (1998 RA2014): Water-to-Water and Brine-to-Water Heat Pumps —Testing and Rating for Performance
Table C403.3.2(2)

210/240—2016: Performance Rating of Unitary Air-conditioning and Air-source Heat Pump Equipment
Table C403.3.2(1), Table C403.3.2(2)

310/380—2014 (CSA-C744-04): Standard for Packaged Terminal Air Conditioners and Heat Pumps
Table C403.3.2(3)

340/360—2015: Performance Rating of Commercial and Industrial Unitary Air-conditioning and Heat Pump Equipment
Table C403.3.2(1), Table C403.3.2(2)

365(I-P)—2009: Commercial and Industrial Unitary Air-conditioning Condensing Units
Table C403.3.2(1), Table C403.3.2(6)

390 (I-P)—2015: Performance Rating of Single Package Vertical Air-conditioners and Heat Pumps
Table C403.3.2(3)

400 (I-P)—2015: Performance Rating of Liquid to Liquid Heat Exchangers
Table C403.3.2(10)

440—2008: Performance Rating of Room Fan Coils—with Addendum 1
C403.11.3

460—2005: Performance Rating of Remote Mechanical-draft Air-cooled Refrigerant Condensers
Table C403.3.2(8)

AHRI—continued

550/590 (I-P)—2015: Performance Rating of Water-chilling and Heat Pump Water-heating Packages Using the Vapor Compression Cycle
C403.3.2.1, Table C403.3.2(7)

560—00: Absorption Water Chilling and Water Heating Packages
Table C403.3.2(7)

1160 (I-P) —2014: Performance Rating of Heat Pump Pool Heaters
Table C404.2

1200 (I-P)—2013: Performance Rating of Commercial Refrigerated Display Merchandisers and Storage Cabinets
C403.10, Table C403.10.1(1), Table C403.10.1(2)

AMCA

Air Movement and Control Association International
30 West University Drive
Arlington Heights, IL 60004-1806

205—12: Energy Efficiency Classification for Fans
C403.8.3

220—08 (R2012): Laboratory Methods of Testing Air Curtain Units for Aerodynamic Performance Rating
C402.5.6

500D—12: Laboratory Methods for Testing Dampers for Rating
C403.7.7

ANSI

American National Standards Institute
25 West 43rd Street, 4th Floor
New York, NY 10036

Z21.10.3/CSA 4.3—11: Gas Water Heaters, Volume III—Storage Water Heaters with Input Ratings Above 75,000 Btu per Hour, Circulating Tank and Instantaneous
Table C404.2

Z21.47/CSA 2.3—12: Gas-fired Central Furnaces
Table C403.3.2(4)

Z83.8/CSA 2.6—09: Gas Unit Heaters, Gas Packaged Heaters, Gas Utility Heaters and Gas-fired Duct Furnaces
Table C403.3.2(4)

APSP

The Association of Pool & Spa Professionals
2111 Eisenhower Avenue, Suite 580
Alexandria, VA 22314

14—2014: American National Standard for Portable Electric Spa Energy Efficiency
C404.8

ASHRAE

ASHRAE
1791 Tullie Circle NE
Atlanta, GA 30329

ASHRAE 127-2007: Method of Testing for Rating Computer
Table C403.3.2(9)

ANSI/ASHRAE/ACCA Standard 183—2007 (RA2014): Peak Cooling and Heating Load Calculations in Buildings, Except Low-rise Residential Buildings
C403.1.1

ASHRAE—2016: ASHRAE HVAC Systems and Equipment Handbook
C403.1.1

ISO/AHRI/ASHRAE 13256-1 (1998 RA2014): Water-to-Air and Brine-to-Air Heat Pumps—Testing and Rating for Performance
Table C403.3.2(2)

ISO/AHRI/ASHRAE 13256-2 (1998 RA2014): Water-to-Water and Brine-to-Water Heat Pumps—Testing and Rating for Performance
Table C403.3.2(2)

ASHRAE—continued

55—2013: Thermal Environmental Conditions for Human Occupancy
Table C407.5.1

90.1—2016: Energy Standard for Buildings Except Low-rise Residential Buildings
C401.2, Table C402.1.3, Table C402.1.4, C406.2, Table C407.6.1, C502.1, C503.1, C504.1

140—2014: Standard Method of Test for the Evaluation of Building Energy Analysis Computer Programs
C407.6.1

146—2011: Testing and Rating Pool Heaters
Table C404.2

ASME

American Society of Mechanical Engineers
Two Park Avenue
New York, NY 10016-5990

ASME A17.1—2016/CSA B44—16: Safety Code for Elevators and Escalators
C405.8.2

ASTM

ASTM International
100 Barr Harbor Drive, P.O. Box C700
West Conshohocken, PA 19428-2959

C90—14: Specification for Load-bearing Concrete Masonry Units
Table C401.3

C1363—11: Standard Test Method for Thermal Performance of Building Materials and Envelope Assemblies by Means of a Hot Box Apparatus
C303.1.4.1, Table C402.1.4, 402.2.7

C1371—15: Standard Test Method for Determination of Emittance of Materials Near Room Temperature Using Portable Emissometers
Table C402.3

C1549—09(2014): Standard Test Method for Determination of Solar Reflectance Near Ambient Temperature Using a Portable Solar Reflectometer
Table C402.3

D1003—13: Standard Test Method for Haze and Luminous Transmittance of Transparent Plastics
C402.4.2.2

E283—04(2012): Test Method for Determining the Rate of Air Leakage Through Exterior Windows, Curtain Walls and Doors Under Specified Pressure Differences Across the Specimen
C402.5.1.2.2, Table C402.5.2, C402.5.7

E408—13: Test Methods for Total Normal Emittance of Surfaces Using Inspection-meter Techniques
Table C402.3

E779—10: Standard Test Method for Determining Air Leakage Rate by Fan Pressurization
C402.5

E903—12: Standard Test Method Solar Absorptance, Reflectance and Transmittance of Materials Using Integrating Spheres (Withdrawn 2005)
Table C402.3

E1677—11: Specification for Air Barrier (AB) Material or Systems for Low-rise Framed Building Walls
C402.5.1.2.2

E1827—11: Standard Test Methods for Determining Airtightness of Building Using an Orifice Blower Door
C402.5, C406.9, C606.4

E1918—06(2015): Standard Test Method for Measuring Solar Reflectance of Horizontal or Low-sloped Surfaces in the Field
Table C402.3

E1980—11: Standard Practice for Calculating Solar Reflectance Index of Horizontal and Low-sloped Opaque Surfaces
Table C402.3, C402.3.2

E2178—13: Standard Test Method for Air Permanence of Building Materials
C402.5.1.2.1

ASTM—continued

E2357—11: Standard Test Method for Determining Air Leakage of Air Barriers Assemblies
C402.5.1.2.2

CRRC

Cool Roof Rating Council
449 15th Street, Suite 400
Oakland, CA 94612

ANSI/CRRC-S100—2016: Standard Test Methods for Determining Radiative Properties of Materials
Table C402.3, C402.3.1

CSA

CSA Group
8501 East Pleasant Valley Road
Cleveland, OH 44131-5516

AAMA/WDMA/CSA 101/I.S.2/A440—17: North American Fenestration Standard/Specification for Windows, Doors and Unit Skylights
Table C402.5.2

CSA B55.1—2015: Test Method for Measuring Efficiency and Pressure Loss of Drain Water Heat Recovery Units
C404.8

CSA B55.2—2015: Drain Water Heat Recovery Units
C404.8

CTI

Cooling Technology Institute
P. O. Box 681807
Houston, TX 77268

ATC 105 (00): Acceptance Test Code for Water Cooling Tower
Table C403.3.2(8)

ATC 105S—11: Acceptance Test Code for Closed Circuit Cooling Towers
Table C403.3.2(8)

ATC 106—11: Acceptance Test for Mechanical Draft Evaporative Vapor Condensers
Table C403.3.2(8)

STD 201—11: Standard for Certification of Water Cooling Towers Thermal Performances
Table C403.3.2(8)

CTI STD 201 RS(15): Performance Rating of Evaporative Heat Rejection Equipment
Table C403.3.2(8)

DASMA

Door & Access Systems Manufacturers Association, International
1300 Sumner Avenue
Cleveland, OH 44115-2851

105—2016: Test Method for Thermal Transmittance and Air Infiltration of Garage Doors and Rolling Doors
C303.1.3, Table C402.5.2

DOE

U.S. Department of Energy
c/o Superintendent of Documents
1000 Independence Avenue SW
Washington, DC 20585

10 CFR, Part 430—2015: Energy Conservation Program for Consumer Products: Test Procedures and Certification and Enforcement Requirement for Plumbing Products; and Certification and Enforcement Requirements for Residential Appliances; Final Rule
Table C403.3.2(4), Table C403.3.2(5), Table C404.2

DOE—continued

10 CFR, Part 430, Subpart B, Appendix N—(2015): Uniform Test Method for Measuring the Energy Consumption of Furnaces and Boilers
C202

10 CFR, Part 431—2015: Energy Efficiency Program for Certain Commercial and Industrial Equipment: Test Procedures and Efficiency Standards; Final Rules
Table C403.3.2(5), C405.6, Table C405.6, C405.7

10 CFR 431 Subpart B App B: Uniform Test Method for Measuring Nominal Full Load Efficiency of Electric Motors
C403.8.4, Table C405.7(1), Table C405.7(2), Table C405.7(3), C405.7(4)

NAECA 87—(88): National Appliance Energy Conservation Act 1987 [Public Law 100-12 (with Amendments of 1988-P.L. 100-357)]
Table C403.3.2(1), Table C403.3.2(2), Table C403.3.2(4)

ICC

International Code Council, Inc.
500 New Jersey Avenue NW
6th Floor
Washington, DC 20001

IBC—18: International Building Code®
C201.3, C303.2, C402.5.3, C501.4

IFC—18: International Fire Code®
C201.3, C501.4

IFGC—18: International Fuel Gas Code®
C201.3, C501.4

IMC—18: International Mechanical Code®
C403.7.7, C403.2.2, C403.7.1, C403.7.2, C403.7.4, C403.7.5, C403.11.1, C403.11.2.1, C403.11.2.2, C403.6, C403.6.6, C406.6, C501.4

IPC—18: International Plumbing Code®
C201.3, C501.4

IPMC—18: International Property Maintenance Code®
C501.4

IPSDC—18: International Private Sewage Disposal Code®
C501.4

IEEE

Institute of Electrical and Electronic Engineers
3 Park Avenue, 17th Floor
New York, NY 10016

IEEE 515.1—2012: IEE Standard for the Testing, Design, Installation, and Maintenance of Electrical Resistance Trace Heating for Commercial Applications
C404.6.2

IES

Illuminating Engineering Society
120 Wall Street, 17th Floor
New York, NY 10005-4001

ANSI/ASHRAE/IESNA 90.1—2016: Energy Standard for Buildings, Except Low-rise Residential Buildings
C401.2, Table C402.1.3, Table C402.1.4, C406.2, C502.1, C503.1, C504.1

ISO

International Organization for Standardization
Chemin de Blandonnet 8, CP 401, 1214 Vernier
Geneva, Switzerland

ISO/AHRI/ASHRAE 13256-1(1998 RA2014): Water-to-Air and Brine-to-Air Heat Pumps -Testing and Rating for Performance
Table C403.3.2(2)

ISO/AHRI/ASHRAE 13256-2(1998 RA2014): Water-to-Water and Brine-to-Water Heat Pumps -Testing and Rating for Performance
C403.3.2(2)

NEMA

National Electrical Manufacturers Association
1300 North 17th Street, Suite 900
Rosslyn, VA 22209

MG1—2014: Motors and Generators

C202

NFPA

National Fire Protection Association
1 Batterymarch Park
Quincy, MA 02169-7471

70—17: National Electrical Code

C501.4

NFRC

National Fenestration Rating Council, Inc.
6305 Ivy Lane, Suite 140
Greenbelt, MD 20770

100—2017: Procedure for Determining Fenestration Products *U-factors*

C303.1.3, C402.2.1.1

200—2017: Procedure for Determining Fenestration Product Solar Heat Gain Coefficients and Visible Transmittance at Normal Incidence

C303.1.3, C402.4.1.1

400—2017: Procedure for Determining Fenestration Product Air Leakage

Table C402.5.2

SMACNA

Sheet Metal and Air Conditioning Contractors' National Association, Inc.
4021 Lafayette Center Drive
Chantilly, VA 20151-1219

SMACNA—2012: HVAC Air Duct Leakage Test Manual Second Edition

C403.2.11.2.3

UL

UL LLC
333 Pfingsten Road
Northbrook, IL 60062-2096

710—12: Exhaust Hoods for Commercial Cooking Equipment—with Revisions through November 2013

C403.7.5

727—06: Oil-fired Central Furnaces—with Revisions through October 2013

Table C403.3.2(4)

731—95: Oil-fired Unit Heaters—with Revisions through October 2013

Table C403.3.2(4)

1784—01: Air Leakage Tests of Door Assemblies—with Revisions through February 2015

C402.5.3

US-FTC

United States-Federal Trade Commission
600 Pennsylvania Avenue NW
Washington, DC 20580

CFR Title 16 (2015): *R*-value Rule

C303.1.4

WDMA

Window and Door Manufacturers Association
2025 M Street NW, Suite 800
Washington, DC 20036-3309

AAMA/WDMA/CSA 101/I.S.2/A440—17: North American Fenestration Standard/Specification for Windows, Doors and Unit Skylights
Table C402.5.2

APPENDIX CA

SOLAR-READY ZONE—COMMERCIAL

The provisions contained in this appendix are not mandatory unless specifically referenced in the adopting ordinance.

User note:

About this appendix: *Appendix CA is intended to encourage the installation of renewable energy systems by preparing buildings for the future installation of solar energy equipment, piping and wiring.*

SECTION CA101
SCOPE

CA101.1 General. These provisions shall be applicable for new construction where solar-ready provisions are required.

SECTION CA102
GENERAL DEFINITION

SOLAR-READY ZONE. A section or sections of the roof or building overhang designated and reserved for the future installation of a solar photovoltaic or solar thermal system.

SECTION CA103
SOLAR-READY ZONE

CA103.1 General. A solar-ready zone shall be located on the roof of buildings that are five stories or less in height above grade plane, and are oriented between 110 degrees and 270 degrees of true north or have low-slope roofs. Solar-ready zones shall comply with Sections CA103.2 through CA103.8.

Exceptions:

1. A building with a permanently installed, on-site renewable energy system.
2. A building with a solar-ready zone that is shaded for more than 70 percent of daylight hours annually.
3. A building where the licensed design professional certifies that the incident solar radiation available to the building is not suitable for a solar-ready zone.
4. A building where the licensed design professional certifies that the solar zone area required by Section CA103.3 cannot be met because of extensive rooftop equipment, skylights, vegetative roof areas or other obstructions.

CA103.2 Construction document requirements for a solar-ready zone. Construction documents shall indicate the solar-ready zone.

CA103.3 Solar-ready zone area. The total solar-ready zone area shall be not less than 40 percent of the roof area calculated as the horizontally projected gross roof area less the area covered by skylights, occupied roof decks, vegetative roof areas and mandatory *access* or set back areas as required by the *International Fire Code*. The solar-ready zone shall be a single area or smaller, separated sub-zone areas. Each subzone shall be not less than 5 feet (1524 mm) in width in the narrowest dimension.

CA103.4 Obstructions. Solar ready zones shall be free from obstructions, including pipes, vents, ducts, HVAC equipment, skylights and roof-mounted equipment.

CA103.5 Roof loads and documentation. A collateral dead load of not less than 5 pounds per square foot (5 psf) (24.41 kg/m^2) shall be included in the gravity and lateral design calculations for the solar-ready zone. The structural design loads for roof dead load and roof live load shall be indicated on the construction documents.

CA103.6 Interconnection pathway. Construction documents shall indicate pathways for routing of conduit or piping from the solar-ready zone to the electrical service panel or service hot water system.

CA103.7 Electrical service reserved space. The main electrical service panel shall have a reserved space to allow installation of a dual-pole circuit breaker for future solar electric installation and shall be labeled "For Future Solar Electric." The reserved space shall be positioned at the end of the panel that is opposite from the panel supply conductor connection.

CA103.8 Construction documentation certificate. A permanent certificate, indicating the solar-ready zone and other requirements of this section, shall be posted near the electrical distribution panel, water heater or other conspicuous location by the builder or registered design professional.

INDEX

D

E

F

G

H

I

S

T

U

V

W

Z

IECC—RESIDENTIAL PROVISIONS

TABLE OF CONTENTS

CHAPTER 1 [RE]

SCOPE AND ADMINISTRATION

User note:

About this chapter: *Chapter 1 establishes the limits of applicability of this code and describes how the code is to be applied and enforced. Chapter 1 is in two parts: Part 1—Scope and Application (Sections 101–102) and Part 2—Administration and Enforcement (Sections 103–109). Section 101 identifies which buildings and structures come under its purview and references other I-Codes as applicable. Standards and codes are scoped to the extent referenced (see Section 107.1).*

This code is intended to be adopted as a legally enforceable document, and it cannot be effective without adequate provisions for its administration and enforcement. The provisions of Chapter 1 establish the authority and duties of the code official appointed by the authority having jurisdiction and also establish the rights and privileges of the design professional, contractor and property owner.

PART 1—SCOPE AND APPLICATION

SECTION R101 SCOPE AND GENERAL REQUIREMENTS

R101.1 Title. This code shall be known as the *Energy Conservation Code* of **[NAME OF JURISDICTION]**, and shall be cited as such. It is referred to herein as "this code."

R101.2 Scope. This code applies to *residential buildings* and the *building* sites and associated systems and equipment.

R101.3 Intent. This code shall regulate the design and construction of *buildings* for the effective use and conservation of energy over the useful life of each building. This code is intended to provide flexibility to permit the use of innovative approaches and techniques to achieve this objective. This code is not intended to abridge safety, health or environmental requirements contained in other applicable codes or ordinances.

R101.4 Applicability. Where, in any specific case, different sections of this code specify different materials, methods of construction or other requirements, the most restrictive shall govern. Where there is a conflict between a general requirement and a specific requirement, the specific requirement shall govern.

R101.4.1 Mixed residential and commercial buildings. Where a *building* includes both *residential* building and *commercial building* portions, each portion shall be separately considered and meet the applicable provisions of the IECC—Commercial Provisions or IECC—Residential Provisions.

R101.5 Compliance. *Residential buildings* shall meet the provisions of IECC—Residential Provisions. *Commercial buildings* shall meet the provisions of IECC—Commercial Provisions.

R101.5.1 Compliance materials. The *code official* shall be permitted to approve specific computer software, worksheets, compliance manuals and other similar materials that meet the intent of this code.

SECTION R102 ALTERNATIVE MATERIALS, DESIGN AND METHODS OF CONSTRUCTION AND EQUIPMENT

R102.1 General. The provisions of this code are not intended to prevent the installation of any material or to prohibit any design or method of construction not specifically prescribed by this code. The *code official* shall have the authority to approve an alternative material, design or method of construction upon application of the owner or the owner's authorized agent. The code official shall first find that the proposed design is satisfactory and complies with the intent of the provisions of this code, and that the material, method or work offered is, for the purpose intended, not less than the equivalent of that prescribed in this code for strength, effectiveness, fire resistance, durability and safety. Where the alternative material, design or method of construction is not *approved*, the *code official* shall respond to the applicant, in writing, stating the reasons why the alternative was not *approved*.

R102.1.1 Above code programs. The *code official* or other authority having jurisdiction shall be permitted to deem a national, state or local energy-efficiency program to exceed the energy efficiency required by this code. *Buildings approved* in writing by such an energy-efficiency program shall be considered to be in compliance with this code. The requirements identified as "mandatory" in Chapter 4 shall be met.

PART 2—ADMINISTRATION AND ENFORCEMENT

SECTION R103 CONSTRUCTION DOCUMENTS

R103.1 General. Construction documents, technical reports and other supporting data shall be submitted in one or more sets with each application for a permit. The construction documents and technical reports shall be prepared by a registered design professional where required by the statutes of the jurisdiction in which the project is to be constructed. Where special conditions exist, the *code official* is authorized to require necessary construction documents to be prepared by a registered design professional.

Exception: The *code official* is authorized to waive the requirements for construction documents or other supporting data if the *code official* determines they are not necessary to confirm compliance with this code.

R103.2 Information on construction documents. Construction documents shall be drawn to scale on suitable material. Electronic media documents are permitted to be submitted where *approved* by the *code official*. Construction documents shall be of sufficient clarity to indicate the location, nature and extent of the work proposed, and show in sufficient detail pertinent data and features of the *building*, systems and equipment as herein governed. Details shall include the following as applicable:

1. Insulation materials and their *R*-values.
2. Fenestration *U*-factors and *solar heat gain coefficients* (SHGC).
3. Area-weighted *U*-factor and *solar heat gain coefficients* (SHGC) calculations.
4. Mechanical system design criteria.
5. Mechanical and service water-heating systems and equipment types, sizes and efficiencies.
6. Equipment and system controls.
7. Duct sealing, duct and pipe insulation and location.
8. Air sealing details.

R103.2.1 Building thermal envelope depiction. The *building thermal envelope* shall be represented on the construction documents.

R103.3 Examination of documents. The *code official* shall examine or cause to be examined the accompanying construction documents and shall ascertain whether the construction indicated and described is in accordance with the requirements of this code and other pertinent laws or ordinances. The *code official* is authorized to utilize a registered design professional, or other *approved* entity not affiliated with the building design or construction, in conducting the review of the plans and specifications for compliance with the code.

R103.3.1 Approval of construction documents. When the *code official* issues a permit where construction documents are required, the construction documents shall be endorsed in writing and stamped "Reviewed for Code Compliance." Such *approved* construction documents shall not be changed, modified or altered without authorization from the *code official*. Work shall be done in accordance with the *approved* construction documents.

One set of construction documents so reviewed shall be retained by the *code official*. The other set shall be returned to the applicant, kept at the site of work and shall be open to inspection by the *code official* or a duly authorized representative.

R103.3.2 Previous approvals. This code shall not require changes in the construction documents, construction or designated occupancy of a structure for which a lawful permit has been heretofore issued or otherwise lawfully authorized, and the construction of which has been pursued in good faith within 180 days after the effective date of this code and has not been abandoned.

R103.3.3 Phased approval. The *code official* shall have the authority to issue a permit for the construction of part of an energy conservation system before the construction documents for the entire system have been submitted or *approved*, provided adequate information and detailed statements have been filed complying with all pertinent requirements of this code. The holders of such permit shall proceed at their own risk without assurance that the permit for the entire energy conservation system will be granted.

R103.4 Amended construction documents. Work shall be installed in accordance with the *approved* construction documents, and any changes made during construction that are not in compliance with the *approved* construction documents shall be resubmitted for approval as an amended set of construction documents.

R103.5 Retention of construction documents. One set of *approved* construction documents shall be retained by the *code official* for a period of not less than 180 days from date of completion of the permitted work, or as required by state or local laws.

SECTION R104
FEES

R104.1 Fees. A permit shall not be issued until the fees prescribed in Section R104.2 have been paid, nor shall an amendment to a permit be released until the additional fee, if any, has been paid.

R104.2 Schedule of permit fees. A fee for each permit shall be paid as required, in accordance with the schedule as established by the applicable governing authority.

R104.3 Work commencing before permit issuance. Any person who commences any work before obtaining the necessary permits shall be subject to an additional fee established by the *code official* that shall be in addition to the required permit fees.

R104.4 Related fees. The payment of the fee for the construction, *alteration*, removal or demolition of work done in connection to or concurrently with the work or activity authorized by a permit shall not relieve the applicant or holder of the permit from the payment of other fees that are prescribed by law.

R104.5 Refunds. The *code official* is authorized to establish a refund policy.

SECTION R105
INSPECTIONS

R105.1 General. Construction or work for which a permit is required shall be subject to inspection by the *code official* or his or her designated agent, and such construction or work shall remain visible and able to be accessed for inspection purposes until *approved*. It shall be the duty of the permit applicant to cause the work to remain visible and able to be accessed for inspection purposes. Neither the *code official* nor the jurisdiction shall be liable for expense entailed in the removal or replacement of any material, product, system or building component required to allow inspection to validate compliance with this code.

R105.2 Required inspections. The *code official* or his or her designated agent, upon notification, shall make the inspections set forth in Sections R105.2.1 through R105.2.5.

R105.2.1 Footing and foundation inspection. Inspections associated with footings and foundations shall verify compliance with the code as to *R-value*, location, thickness, depth of burial and protection of insulation as required by the code and *approved* plans and specifications.

R105.2.2 Framing and rough-in inspection. Inspections at framing and rough-in shall be made before application of interior finish and shall verify compliance with the code as to: types of insulation and corresponding *R-values* and their correct location and proper installation; fenestration properties such as *U*-factor and SHGC and proper installation; and air leakage controls as required by the code; and approved plans and specifications.

R105.2.3 Plumbing rough-in inspection. Inspections at plumbing rough-in shall verify compliance as required by the code and *approved* plans and specifications as to types of insulation and corresponding *R*-values and protection, and required controls.

R105.2.4 Mechanical rough-in inspection. Inspections at mechanical rough-in shall verify compliance as required by the code and *approved* plans and specifications as to installed HVAC equipment type and size, required controls, system insulation and corresponding *R*-value, system air leakage control, programmable thermostats, dampers, whole-house ventilation, and minimum fan efficiency.

Exception: Systems serving multiple dwelling units shall be inspected in accordance with Section C105.2.4.

R105.2.5 Final inspection. The *building* shall have a final inspection and shall not be occupied until *approved*. The final inspection shall include verification of the installation of all required *building* systems, equipment and controls and their proper operation and the required number of high-efficacy lamps and fixtures.

R105.3 Reinspection. A *building* shall be reinspected where determined necessary by the *code official*.

R105.4 Approved inspection agencies. The *code official* is authorized to accept reports of third-party inspection agencies not affiliated with the *building* design or construction, provided that such agencies are *approved* as to qualifications and reliability relevant to the *building* components and systems that they are inspecting.

R105.5 Inspection requests. It shall be the duty of the holder of the permit or their duly authorized agent to notify the *code official* when work is ready for inspection. It shall be the duty of the permit holder to provide access to and means for inspections of such work that are required by this code.

R105.6 Reinspection and testing. Where any work or installation does not pass an initial test or inspection, the necessary corrections shall be made to achieve compliance with this code. The work or installation shall then be resubmitted to the *code official* for inspection and testing.

R105.7 Approval. After the prescribed tests and inspections indicate that the work complies in all respects with this code, a notice of approval shall be issued by the *code official*.

R105.7.1 Revocation. The *code official* is authorized to, in writing, suspend or revoke a notice of approval issued under the provisions of this code wherever the certificate is issued in error, or on the basis of incorrect information supplied, or where it is determined that the *building* or structure, premise, or portion thereof is in violation of any ordinance or regulation or any of the provisions of this code.

SECTION R106
VALIDITY

R106.1 General. If a portion of this code is held to be illegal or void, such a decision shall not affect the validity of the remainder of this code.

SECTION R107
REFERENCED STANDARDS

R107.1 Referenced codes and standards. The codes and standards referenced in this code shall be those indicated in Chapter 5, and such codes and standards shall be considered as part of the requirements of this code to the prescribed extent of each such reference and as further regulated in Sections R107.1.1 and R107.1.2.

R107.1.1 Conflicts. Where conflicts occur between provisions of this code and referenced codes and standards, the provisions of this code shall apply.

R107.1.2 Provisions in referenced codes and standards. Where the extent of the reference to a referenced code or standard includes subject matter that is within the scope of this code, the provisions of this code, as applicable, shall take precedence over the provisions in the referenced code or standard.

R107.2 Application of references. References to chapter or section numbers, or to provisions not specifically identified by number, shall be construed to refer to such chapter, section or provision of this code.

R107.3 Other laws. The provisions of this code shall not be deemed to nullify any provisions of local, state or federal law.

SECTION R108
STOP WORK ORDER

R108.1 Authority. Where the *code official* finds any work regulated by this code being performed in a manner either contrary to the provisions of this code or dangerous or unsafe, the *code official* is authorized to issue a stop work order.

R108.2 Issuance. The stop work order shall be in writing and shall be given to the owner of the property involved, to the owner's authorized agent, or to the person doing the work. Upon issuance of a stop work order, the cited work shall immediately cease. The stop work order shall state the reason for the order and the conditions under which the cited work will be permitted to resume.

R108.3 Emergencies. Where an emergency exists, the *code official* shall not be required to give a written notice prior to stopping the work.

R108.4 Failure to comply. Any person who shall continue any work after having been served with a stop work order, except such work as that person is directed to perform to remove a violation or unsafe condition, shall be subject to a fine as set by the applicable governing authority.

SECTION R109 BOARD OF APPEALS

R109.1 General. In order to hear and decide appeals of orders, decisions or determinations made by the *code official* relative to the application and interpretation of this code, there shall be and is hereby created a board of appeals. The *code official* shall be an ex officio member of said board but shall not have a vote on any matter before the board. The board of appeals shall be appointed by the governing body and shall hold office at its pleasure. The board shall adopt rules of procedure for conducting its business, and shall render all decisions and findings in writing to the appellant with a duplicate copy to the *code official*.

R109.2 Limitations on authority. An application for appeal shall be based on a claim that the true intent of this code or the rules legally adopted thereunder have been incorrectly interpreted, the provisions of this code do not fully apply or an equally good or better form of construction is proposed. The board shall not have authority to waive requirements of this code.

R109.3 Qualifications. The board of appeals shall consist of members who are qualified by experience and training and are not employees of the jurisdiction.

CHAPTER 2 [RE]

DEFINITIONS

User note:

About this chapter: *Codes, by their very nature, are technical documents. Every word, term and punctuation mark can add to or change the meaning of a technical requirement. It is necessary to maintain a consensus on the specific meaning of each term contained in the code. Chapter 2 performs this function by stating clearly what specific terms mean for the purpose of the code.*

SECTION R201
GENERAL

R201.1 Scope. Unless stated otherwise, the following words and terms in this code shall have the meanings indicated in this chapter.

R201.2 Interchangeability. Words used in the present tense include the future; words in the masculine gender include the feminine and neuter; the singular number includes the plural and the plural includes the singular.

R201.3 Terms defined in other codes. Terms that are not defined in this code but are defined in the *International Building Code*, *International Fire Code*, *International Fuel Gas Code*, *International Mechanical Code*, *International Plumbing Code* or the *International Residential Code* shall have the meanings ascribed to them in those codes.

R201.4 Terms not defined. Terms not defined by this chapter shall have ordinarily accepted meanings such as the context implies.

SECTION R202
GENERAL DEFINITIONS

ABOVE-GRADE WALL. A wall more than 50 percent above grade and enclosing *conditioned space*. This includes between-floor spandrels, peripheral edges of floors, roof and basement knee walls, dormer walls, gable end walls, walls enclosing a mansard roof and *skylight* shafts.

ACCESSIBLE. Admitting close approach as a result of not being guarded by locked doors, elevation or other effective means (see "Readily *accessible*").

ADDITION. An extension or increase in the *conditioned space* floor area, number of stories or height of a building or structure.

AIR BARRIER. One or more materials joined together in a continuous manner to restrict or prevent the passage of air through the *building thermal envelope* and its assemblies.

AIR-IMPERMEABLE INSULATION. An insulation that functions as an air barrier material.

ALTERATION. Any construction, retrofit or renovation to an existing structure other than *repair* or *addition*. Also, a change in a building, electrical, gas, mechanical or plumbing system that involves an extension, addition or change to the arrangement, type or purpose of the original installation.

APPROVED. Acceptable to the *code official*.

APPROVED AGENCY. An established and recognized agency that is regularly engaged in conducting tests furnishing inspection services, or furnishing product certification, where such agency has been *approved* by the *code official*.

AUTOMATIC. Self-acting, operating by its own mechanism when actuated by some impersonal influence, as, for example, a change in current strength, pressure, temperature or mechanical configuration (see "Manual").

BASEMENT WALL. A wall 50 percent or more below grade and enclosing *conditioned space*.

BUILDING. Any structure used or intended for supporting or sheltering any use or occupancy, including any mechanical systems, service water heating systems and electric power and lighting systems located on the building site and supporting the building.

BUILDING SITE. A contiguous area of land that is under the ownership or control of one entity.

BUILDING THERMAL ENVELOPE. The *basement walls, exterior walls*, floors, ceiling, roofs and any other *building* element assemblies that enclose *conditioned space* or provide a boundary between *conditioned space* and exempt or unconditioned space.

CIRCULATING HOT WATER SYSTEM. A specifically designed water distribution system where one or more pumps are operated in the service hot water piping to circulate heated water from the water-heating equipment to fixtures and back to the water-heating equipment.

CLIMATE ZONE. A geographical region based on climatic criteria as specified in this code.

CODE OFFICIAL. The officer or other designated authority charged with the administration and enforcement of this code, or a duly authorized representative.

COMMERCIAL BUILDING. For this code, all buildings that are not included in the definition of "Residential building."

CONDITIONED FLOOR AREA. The horizontal projection of the floors associated with the *conditioned space*.

CONDITIONED SPACE. An area, room or space that is enclosed within the *building thermal envelope* and that is directly or indirectly heated or cooled. Spaces are indirectly heated or cooled where they communicate through openings with conditioned spaces, where they are separated from conditioned spaces by uninsulated walls, floors or ceilings, or where they contain uninsulated ducts, piping or other sources of heating or cooling.

CONTINUOUS AIR BARRIER. A combination of materials and assemblies that restrict or prevent the passage of air through the *building thermal envelope*.

CONTINUOUS INSULATION (ci). Insulating material that is continuous across all structural members without thermal bridges other than fasteners and service openings. It is installed on the interior or exterior, or is integral to any opaque surface, of the *building* envelope.

CRAWL SPACE WALL. The opaque portion of a wall that encloses a crawl space and is partially or totally below grade.

CURTAIN WALL. Fenestration products used to create an external nonload-bearing wall that is designed to separate the exterior and interior environments.

DEMAND RECIRCULATION WATER SYSTEM. A water distribution system having one or more recirculation pumps that pump water from a heated water supply pipe back to the heated water source through a cold water supply pipe.

DUCT. A tube or conduit utilized for conveying air. The air passages of self-contained systems are not to be construed as air ducts.

DUCT SYSTEM. A continuous passageway for the transmission of air that, in addition to ducts, includes duct fittings, dampers, plenums, fans and accessory air-handling equipment and appliances.

DWELLING UNIT. A single unit providing complete independent living facilities for one or more persons, including permanent provisions for living, sleeping, eating, cooking and sanitation.

ENERGY ANALYSIS. A method for estimating the annual energy use of the *proposed design* and *standard reference design* based on estimates of energy use.

ENERGY COST. The total estimated annual cost for purchased energy for the building functions regulated by this code, including applicable demand charges.

ENERGY SIMULATION TOOL. An *approved* software program or calculation-based methodology that projects the annual energy use of a *building*.

ERI REFERENCE DESIGN. A version of the *rated design* that meets the minimum requirements of the 2006 *International Energy Conservation Code*.

EXTERIOR WALL. Walls including both above-grade walls and *basement walls*.

FENESTRATION. Products classified as either *vertical fenestration* or *skylights*.

Skylights. Glass or other transparent or translucent glazing material installed at a slope of less than 60 degrees (1.05 rad) from horizontal.

Vertical fenestration. Windows that are fixed or operable, opaque doors, glazed doors, glazed block and combination opaque/glazed doors composed of glass or other transparent or translucent glazing materials and installed at a slope of not less than 60 degrees (1.05 rad) from horizontal.

FENESTRATION PRODUCT, SITE-BUILT. A fenestration designed to be made up of field-glazed or field-assembled units using specific factory cut or otherwise factory-formed framing and glazing units. Examples of site-built fenestration include storefront systems, curtain walls and atrium roof systems.

HEATED SLAB. Slab-on-grade construction in which the heating elements, hydronic tubing, or hot air distribution system is in contact with, or placed within or under, the slab.

HIGH-EFFICACY LAMPS. Compact fluorescent lamps, light-emitting diode (LED) lamps, T-8 or smaller diameter linear fluorescent lamps, or other lamps with an efficacy of not less than the following:

1. 60 lumens per watt for lamps over 40 watts.
2. 50 lumens per watt for lamps over 15 watts to 40 watts.
3. 40 lumens per watt for lamps 15 watts or less.

HISTORIC BUILDING. Any building or structure that is one or more of the following:

1. Listed, or certified as eligible for listing by the State Historic Preservation Officer or the Keeper of the National Register of Historic Places, in the National Register of Historic Places.
2. Designated as historic under an applicable state or local law.
3. Certified as a contributing resource within a National Register-listed, state-designated or locally designated historic district.

INFILTRATION. The uncontrolled inward air leakage into a *building* caused by the pressure effects of wind or the effect of differences in the indoor and outdoor air density or both.

INSULATED SIDING. A type of continuous insulation with manufacturer-installed insulating material as an integral part of the cladding product having an *R*-value of not less than R-2.

LABELED. Equipment, materials or products to which have been affixed a label, seal, symbol or other identifying mark of a nationally recognized testing laboratory, *approved* agency or other organization concerned with product evaluation that maintains periodic inspection of the production of such labeled items and whose labeling indicates either that the equipment, material or product meets identified standards or has been tested and found suitable for a specified purpose.

LISTED. Equipment, materials, products or services included in a list published by an organization acceptable to the *code official* and concerned with evaluation of products or services that maintains periodic inspection of production of *listed* equipment or materials or periodic evaluation of services and whose listing states either that the equipment, material, product or service meets identified standards or has been tested and found suitable for a specified purpose.

LOW-VOLTAGE LIGHTING. Lighting equipment powered through a transformer such as a cable conductor, a rail conductor and track lighting.

MANUAL. Capable of being operated by personal intervention (see "Automatic").

OPAQUE DOOR. A door that is not less than 50-percent opaque in surface area.

PROPOSED DESIGN. A description of the proposed *building* used to estimate annual energy use for determining compliance based on total building performance.

RATED DESIGN. A description of the proposed *building* used to determine the energy rating index.

READILY ACCESSIBLE. Capable of being reached quickly for operation, renewal or inspection without requiring those to whom ready access is requisite to climb over or remove obstacles or to resort to portable ladders or access equipment (see "*Accessible*").

REPAIR. The reconstruction or renewal of any part of an existing *building* for the purpose of its maintenance or to correct damage.

REROOFING. The process of recovering or replacing an existing roof covering. See "Roof recover" and "Roof replacement."

RESIDENTIAL BUILDING. For this code, includes detached one- and two-family dwellings and townhouses as well as *Group R*-2, R-3 and R-4 buildings three stories or less in height above grade plane.

ROOF ASSEMBLY. A system designed to provide weather protection and resistance to design loads. The system consists of a roof covering and roof deck or a single component serving as both the roof covering and the roof deck. A roof assembly includes the roof covering, underlayment and roof deck, and can also include a thermal barrier, an ignition barrier, insulation or a vapor retarder.

ROOF RE-COVER. The process of installing an additional roof covering over a prepared existing roof covering without removing the existing roof covering.

ROOF REPAIR. Reconstruction or renewal of any part of an existing roof for the purposes of its maintenance.

ROOF REPLACEMENT. The process of removing the existing roof covering, repairing any damaged substrate and installing a new roof covering.

***R*-VALUE (THERMAL RESISTANCE).** The inverse of the time rate of heat flow through a body from one of its bounding surfaces to the other surface for a unit temperature difference between the two surfaces, under steady state conditions, per unit area (h • ft^2 • °F/Btu) [(m^2 • K)/W].

SERVICE WATER HEATING. Supply of hot water for purposes other than comfort heating.

SOLAR HEAT GAIN COEFFICIENT (SHGC). The ratio of the solar heat gain entering the space through the fenestration assembly to the incident solar radiation. Solar heat gain includes directly transmitted solar heat and absorbed solar radiation that is then reradiated, conducted or convected into the space.

STANDARD REFERENCE DESIGN. A version of the *proposed design* that meets the minimum requirements of this code and is used to determine the maximum annual energy use requirement for compliance based on total building performance.

SUNROOM. A one-story structure attached to a dwelling with a glazing area in excess of 40 percent of the gross area of the structure's *exterior walls* and roof.

THERMAL ISOLATION. Physical and space conditioning separation from *conditioned spaces*. The *conditioned spaces* shall be controlled as separate zones for heating and cooling or conditioned by separate equipment.

THERMOSTAT. An automatic control device used to maintain temperature at a fixed or adjustable setpoint.

***U*-FACTOR (THERMAL TRANSMITTANCE).** The coefficient of heat transmission (air to air) through a building component or assembly, equal to the time rate of heat flow per unit area and unit temperature difference between the warm side and cold side air films (Btu/h • ft^2 • °F) [W/(m^2 • K)].

VENTILATION. The natural or mechanical process of supplying conditioned or unconditioned air to, or removing such air from, any space.

VENTILATION AIR. That portion of supply air that comes from outside (outdoors) plus any recirculated air that has been treated to maintain the desired quality of air within a designated space.

VISIBLE TRANSMITTANCE [VT]. The ratio of visible light entering the space through the fenestration product assembly to the incident visible light, Visible Transmittance, includes the effects of glazing material and frame and is expressed as a number between 0 and 1.

WHOLE HOUSE MECHANICAL VENTILATION SYSTEM. An exhaust system, supply system, or combination thereof that is designed to mechanically exchange indoor air with outdoor air when operating continuously or through a programmed intermittent schedule to satisfy the whole house ventilation rates.

ZONE. A space or group of spaces within a *building* with heating or cooling requirements that are sufficiently similar so that desired conditions can be maintained throughout using a single controlling device.

CHAPTER 3 [RE]

GENERAL REQUIREMENTS

User note:

About this chapter: *Chapter 3 covers general regulations for energy conservation features of buildings. The climate zone for a building is established by geographic location tables and figures in this chapter.*

SECTION R301
CLIMATE ZONES

R301.1 General. *Climate zones* from Figure R301.1 or Table R301.1 shall be used for determining the applicable requirements from Chapter 4. Locations not indicated in Table R301.1 shall be assigned a *climate zone* in accordance with Section R301.3.

R301.2 Warm humid counties. In Table R301.1, warm humid counties are identified by an asterisk.

R301.3 International climate zones. The *climate zone* for any location outside the United States shall be determined by applying Table R301.3(1) and then Table R301.3(2).

R301.4 Tropical climate zone. The tropical *climate zone* shall be defined as:

1. Hawaii, Puerto Rico, Guam, American Samoa, U.S. Virgin Islands, Commonwealth of Northern Mariana Islands; and
2. Islands in the area between the Tropic of Cancer and the Tropic of Capricorn.

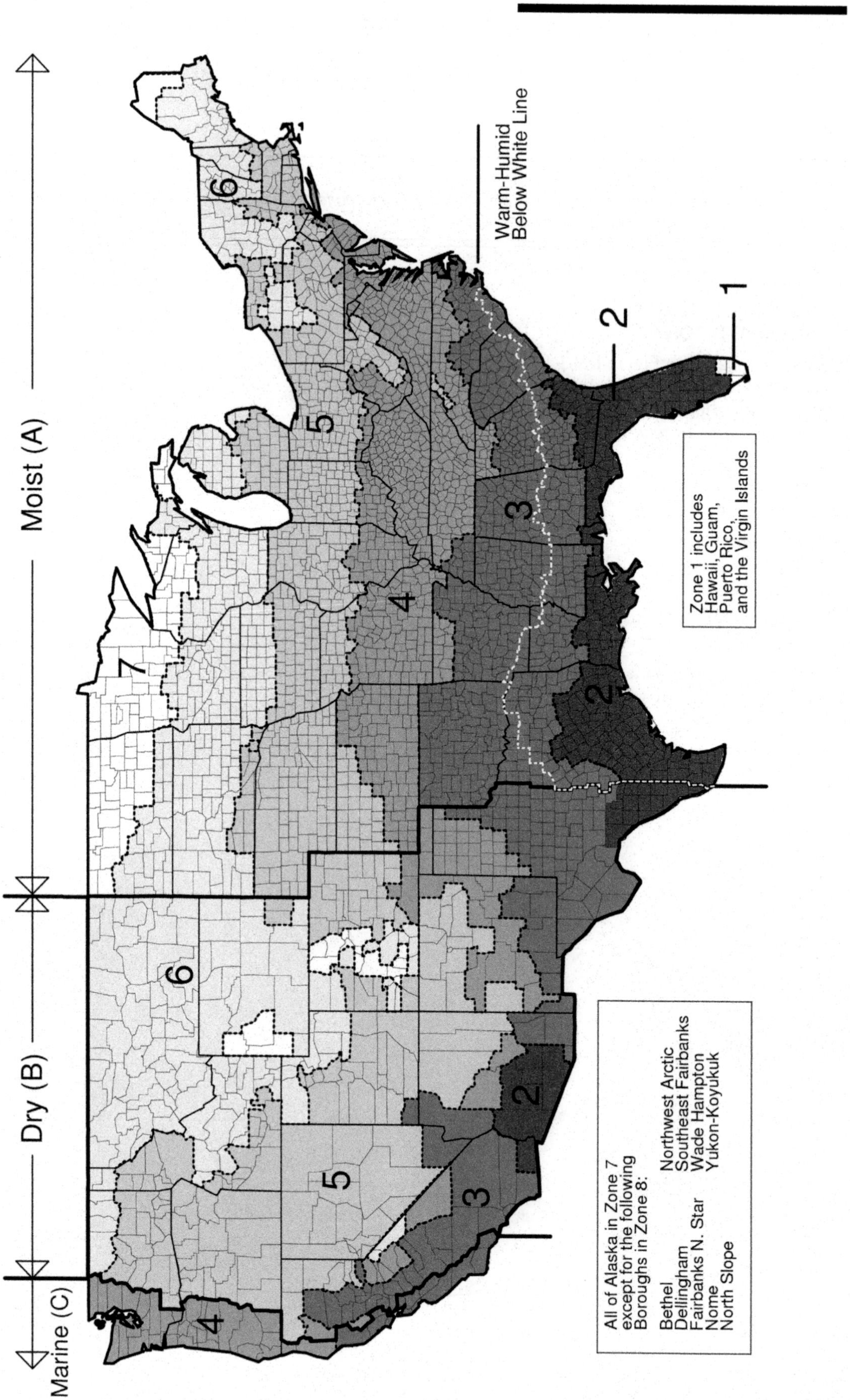

Moist (A)

Dry (B)

Marine (C)

Warm-Humid
Below White Line

7

6

5

4

3

2

1

All of Alaska in Zone 7
except for the following
Boroughs in Zone 8:

Bethel
Dellingham
Fairbanks N. Star
Nome
North Slope
Northwest Arctic
Southeast Fairbanks
Wade Hampton
Yukon-Koyukuk

Zone 1 includes
Hawaii, Guam,
Puerto Rico,
and the Virgin Islands

FIGURE R301.1
CLIMATE ZONES

**TABLE R301.1
CLIMATE ZONES, MOISTURE REGIMES, AND WARM-HUMID
DESIGNATIONS BY STATE, COUNTY AND TERRITORY**

Key: A – Moist, B – Dry, C – Marine. Absence of moisture designation indicates moisture regime is irrelevant. Asterisk (*) indicates a warm-humid location.

US STATES

ALABAMA

3A Autauga*
2A Baldwin*
3A Barbour*
3A Bibb
3A Blount
3A Bullock*
3A Butler*
3A Calhoun
3A Chambers
3A Cherokee
3A Chilton
3A Choctaw*
3A Clarke*
3A Clay
3A Cleburne
3A Coffee*
3A Colbert
3A Conecuh*
3A Coosa
3A Covington*
3A Crenshaw*
3A Cullman
3A Dale*
3A Dallas*
3A DeKalb
3A Elmore*
3A Escambia*
3A Etowah
3A Fayette
3A Franklin
3A Geneva*
3A Greene
3A Hale
3A Henry*
3A Houston*
3A Jackson
3A Jefferson
3A Lamar
3A Lauderdale
3A Lawrence
3A Lee
3A Limestone
3A Lowndes*
3A Macon*
3A Madison
3A Marengo*
3A Marion
3A Marshall
2A Mobile*
3A Monroe*
3A Montgomery*
3A Morgan
3A Perry*
3A Pickens
3A Pike*
3A Randolph
3A Russell*
3A Shelby
3A St. Clair
3A Sumter
3A Talladega
3A Tallapoosa
3A Tuscaloosa
3A Walker
3A Washington*
3A Wilcox*
3A Winston

ALASKA

7 Aleutians East
7 Aleutians West
7 Anchorage
8 Bethel
7 Bristol Bay
7 Denali
8 Dillingham
8 Fairbanks North Star
7 Haines
7 Juneau
7 Kenai Peninsula
7 Ketchikan Gateway
7 Kodiak Island
7 Lake and Peninsula
7 Matanuska-Susitna
8 Nome
8 North Slope
8 Northwest Arctic
7 Prince of Wales-Outer Ketchikan
7 Sitka
7 Skagway-Hoonah-Angoon
8 Southeast Fairbanks
7 Valdez-Cordova
8 Wade Hampton
7 Wrangell-Petersburg
7 Yakutat
8 Yukon-Koyukuk

ARIZONA

5B Apache
3B Cochise
5B Coconino
4B Gila
3B Graham
3B Greenlee
2B La Paz
2B Maricopa
3B Mohave
5B Navajo
2B Pima
2B Pinal
3B Santa Cruz
4B Yavapai
2B Yuma

ARKANSAS

3A Arkansas
3A Ashley
4A Baxter
4A Benton
4A Boone
3A Bradley
3A Calhoun
4A Carroll
3A Chicot
3A Clark
3A Clay
3A Cleburne
3A Cleveland
3A Columbia*
3A Conway
3A Craighead
3A Crawford
3A Crittenden
3A Cross
3A Dallas
3A Desha
3A Drew
3A Faulkner
3A Franklin
4A Fulton
3A Garland
3A Grant
3A Greene
3A Hempstead*
3A Hot Spring
3A Howard
3A Independence
4A Izard
3A Jackson
3A Jefferson
3A Johnson
3A Lafayette*
3A Lawrence
3A Lee
3A Lincoln
3A Little River*
3A Logan
3A Lonoke
4A Madison
4A Marion
3A Miller*
3A Mississippi
3A Monroe
3A Montgomery
3A Nevada
4A Newton
3A Ouachita
3A Perry
3A Phillips
3A Pike
3A Poinsett
3A Polk
3A Pope
3A Prairie
3A Pulaski
3A Randolph
3A Saline
3A Scott
4A Searcy
3A Sebastian
3A Sevier*
3A Sharp
3A St. Francis
4A Stone
3A Union*
3A Van Buren
4A Washington
3A White
3A Woodruff
3A Yell

CALIFORNIA

3C Alameda
6B Alpine
4B Amador
3B Butte
4B Calaveras
3B Colusa
3B Contra Costa
4C Del Norte
4B El Dorado
3B Fresno
3B Glenn

(continued)

TABLE R301.1—continued
CLIMATE ZONES, MOISTURE REGIMES, AND WARM-HUMID DESIGNATIONS BY STATE, COUNTY AND TERRITORY

4C Humboldt
2B Imperial
4B Inyo
3B Kern
3B Kings
4B Lake
5B Lassen
3B Los Angeles
3B Madera
3C Marin
4B Mariposa
3C Mendocino
3B Merced
5B Modoc
6B Mono
3C Monterey
3C Napa
5B Nevada
3B Orange
3B Placer
5B Plumas
3B Riverside
3B Sacramento
3C San Benito
3B San Bernardino
3B San Diego
3C San Francisco
3B San Joaquin
3C San Luis Obispo
3C San Mateo
3C Santa Barbara
3C Santa Clara
3C Santa Cruz
3B Shasta
5B Sierra
5B Siskiyou
3B Solano
3C Sonoma
3B Stanislaus
3B Sutter
3B Tehama
4B Trinity
3B Tulare
4B Tuolumne
3C Ventura
3B Yolo
3B Yuba

COLORADO

5B Adams
6B Alamosa
5B Arapahoe
6B Archuleta
4B Baca
5B Bent
5B Boulder
5B Broomfield
6B Chaffee
5B Cheyenne
7 Clear Creek
6B Conejos
6B Costilla
5B Crowley
6B Custer
5B Delta
5B Denver
6B Dolores
5B Douglas
6B Eagle
5B Elbert
5B El Paso
5B Fremont
5B Garfield
5B Gilpin
7 Grand
7 Gunnison
7 Hinsdale
5B Huerfano
7 Jackson
5B Jefferson
5B Kiowa
5B Kit Carson
7 Lake
5B La Plata
5B Larimer
4B Las Animas
5B Lincoln
5B Logan
5B Mesa
7 Mineral
6B Moffat
5B Montezuma
5B Montrose
5B Morgan
4B Otero
6B Ouray
7 Park
5B Phillips
7 Pitkin
5B Prowers
5B Pueblo
6B Rio Blanco
7 Rio Grande
7 Routt
6B Saguache
7 San Juan
6B San Miguel
5B Sedgwick
7 Summit
5B Teller
5B Washington
5B Weld
5B Yuma

CONNECTICUT

5A (all)

DELAWARE

4A (all)

DISTRICT OF COLUMBIA

4A (all)

FLORIDA

2A Alachua*
2A Baker*
2A Bay*
2A Bradford*
2A Brevard*
1A Broward*
2A Calhoun*
2A Charlotte*
2A Citrus*
2A Clay*
2A Collier*
2A Columbia*
2A DeSoto*
2A Dixie*
2A Duval*
2A Escambia*
2A Flagler*
2A Franklin*
2A Gadsden*
2A Gilchrist*
2A Glades*
2A Gulf*
2A Hamilton*
2A Hardee*
2A Hendry*
2A Hernando*
2A Highlands*
2A Hillsborough*
2A Holmes*
2A Indian River*
2A Jackson*
2A Jefferson*
2A Lafayette*
2A Lake*
2A Lee*
2A Leon*
2A Levy*
2A Liberty*
2A Madison*
2A Manatee*
2A Marion*
2A Martin*
1A Miami-Dade*
1A Monroe*
2A Nassau*
2A Okaloosa*
2A Okeechobee*
2A Orange*
2A Osceola*
2A Palm Beach*
2A Pasco*
2A Pinellas*
2A Polk*
2A Putnam*
2A Santa Rosa*
2A Sarasota*
2A Seminole*
2A St. Johns*
2A St. Lucie*
2A Sumter*
2A Suwannee*
2A Taylor*
2A Union*
2A Volusia*
2A Wakulla*
2A Walton*
2A Washington*

GEORGIA

2A Appling*
2A Atkinson*
2A Bacon*
2A Baker*
3A Baldwin
4A Banks
3A Barrow
3A Bartow
3A Ben Hill*
2A Berrien*
3A Bibb
3A Bleckley*
2A Brantley*
2A Brooks*
2A Bryan*
3A Bulloch*
3A Burke
3A Butts
3A Calhoun*
2A Camden*
3A Candler*
3A Carroll
4A Catoosa
2A Charlton*
2A Chatham*
3A Chattahoochee*
4A Chattooga
3A Cherokee
3A Clarke
3A Clay*
3A Clayton
2A Clinch*
3A Cobb
3A Coffee*
2A Colquitt*
3A Columbia
2A Cook*
3A Coweta

(continued)

TABLE R301.1—continued
CLIMATE ZONES, MOISTURE REGIMES, AND WARM-HUMID DESIGNATIONS BY STATE, COUNTY AND TERRITORY

3A Crawford
3A Crisp*
4A Dade
4A Dawson
2A Decatur*
3A DeKalb
3A Dodge*
3A Dooly*
3A Dougherty*
3A Douglas
3A Early*
2A Echols*
2A Effingham*
3A Elbert
3A Emanuel*
2A Evans*
4A Fannin
3A Fayette
4A Floyd
3A Forsyth
4A Franklin
3A Fulton
4A Gilmer
3A Glascock
2A Glynn*
4A Gordon
2A Grady*
3A Greene
3A Gwinnett
4A Habersham
4A Hall
3A Hancock
3A Haralson
3A Harris
3A Hart
3A Heard
3A Henry
3A Houston*
3A Irwin*
3A Jackson
3A Jasper
2A Jeff Davis*
3A Jefferson
3A Jenkins*
3A Johnson*
3A Jones
3A Lamar
2A Lanier*
3A Laurens*
3A Lee*
2A Liberty*
3A Lincoln
2A Long*
2A Lowndes*
4A Lumpkin
3A Macon*
3A Madison
3A Marion*
3A McDuffie
2A McIntosh*
3A Meriwether
2A Miller*
2A Mitchell*
3A Monroe
3A Montgomery*
3A Morgan
4A Murray
3A Muscogee
3A Newton
3A Oconee
3A Oglethorpe
3A Paulding
3A Peach*
4A Pickens
2A Pierce*
3A Pike
3A Polk
3A Pulaski*
3A Putnam
3A Quitman*
4A Rabun
3A Randolph*
3A Richmond
3A Rockdale
3A Schley*
3A Screven*
2A Seminole*
3A Spalding
4A Stephens
3A Stewart*
3A Sumter*
3A Talbot
3A Taliaferro
2A Tattnall*
3A Taylor*
3A Telfair*
3A Terrell*
2A Thomas*
3A Tift*
2A Toombs*
4A Towns
3A Treutlen*
3A Troup
3A Turner*
3A Twiggs*
4A Union
3A Upson
4A Walker
3A Walton
2A Ware*
3A Warren
3A Washington
2A Wayne*
3A Webster*
3A Wheeler*
4A White
4A Whitfield
3A Wilcox*
3A Wilkes
3A Wilkinson
3A Worth*

HAWAII

1A (all)*

IDAHO

5B Ada
6B Adams
6B Bannock
6B Bear Lake
5B Benewah
6B Bingham
6B Blaine
6B Boise
6B Bonner
6B Bonneville
6B Boundary
6B Butte
6B Camas
5B Canyon
6B Caribou
5B Cassia
6B Clark
5B Clearwater
6B Custer
5B Elmore
6B Franklin
6B Fremont
5B Gem
5B Gooding
5B Idaho
6B Jefferson
5B Jerome
5B Kootenai
5B Latah
6B Lemhi
5B Lewis
5B Lincoln
6B Madison
5B Minidoka
5B Nez Perce
6B Oneida
5B Owyhee
5B Payette
5B Power
5B Shoshone
6B Teton
5B Twin Falls
6B Valley
5B Washington

ILLINOIS

5A Adams
4A Alexander
4A Bond
5A Boone
5A Brown
5A Bureau
5A Calhoun
5A Carroll
5A Cass
5A Champaign
4A Christian
5A Clark
4A Clay
4A Clinton
5A Coles
5A Cook
4A Crawford
5A Cumberland
5A DeKalb
5A De Witt
5A Douglas
5A DuPage
5A Edgar
4A Edwards
4A Effingham
4A Fayette
5A Ford
4A Franklin
5A Fulton
4A Gallatin
5A Greene
5A Grundy
4A Hamilton
5A Hancock
4A Hardin
5A Henderson
5A Henry
5A Iroquois
4A Jackson
4A Jasper
4A Jefferson
5A Jersey
5A Jo Daviess
4A Johnson
5A Kane
5A Kankakee
5A Kendall
5A Knox
5A Lake
5A La Salle
4A Lawrence
5A Lee
5A Livingston
5A Logan
5A Macon
4A Macoupin
4A Madison
4A Marion
5A Marshall
5A Mason
4A Massac
5A McDonough
5A McHenry

(continued)

**TABLE R301.1—continued
CLIMATE ZONES, MOISTURE REGIMES, AND WARM-HUMID
DESIGNATIONS BY STATE, COUNTY AND TERRITORY**

5A McLean
5A Menard
5A Mercer
4A Monroe
4A Montgomery
5A Morgan
5A Moultrie
5A Ogle
5A Peoria
4A Perry
5A Piatt
5A Pike
4A Pope
4A Pulaski
5A Putnam
4A Randolph
4A Richland
5A Rock Island
4A Saline
5A Sangamon
5A Schuyler
5A Scott
4A Shelby
5A Stark
4A St. Clair
5A Stephenson
5A Tazewell
4A Union
5A Vermilion
4A Wabash
5A Warren
4A Washington
4A Wayne
4A White
5A Whiteside
5A Will
4A Williamson
5A Winnebago
5A Woodford

INDIANA

5A Adams
5A Allen
5A Bartholomew
5A Benton
5A Blackford
5A Boone
4A Brown
5A Carroll
5A Cass
4A Clark
5A Clay
5A Clinton
4A Crawford
4A Daviess
4A Dearborn
5A Decatur
5A De Kalb
5A Delaware
4A Dubois
5A Elkhart
5A Fayette
4A Floyd
5A Fountain
5A Franklin
5A Fulton
4A Gibson
5A Grant
4A Greene
5A Hamilton
5A Hancock
4A Harrison
5A Hendricks
5A Henry
5A Howard
5A Huntington
4A Jackson
5A Jasper
5A Jay
4A Jefferson
4A Jennings
5A Johnson
4A Knox
5A Kosciusko
5A LaGrange
5A Lake
5A LaPorte
4A Lawrence
5A Madison
5A Marion
5A Marshall
4A Martin
5A Miami
4A Monroe
5A Montgomery
5A Morgan
5A Newton
5A Noble
4A Ohio
4A Orange
5A Owen
5A Parke
4A Perry
4A Pike
5A Porter
4A Posey
5A Pulaski
5A Putnam
5A Randolph
4A Ripley
5A Rush
4A Scott
5A Shelby
4A Spencer
5A Starke
5A Steuben
5A St. Joseph
4A Sullivan
4A Switzerland
5A Tippecanoe
5A Tipton
5A Union
4A Vanderburgh
5A Vermillion
5A Vigo
5A Wabash
5A Warren
4A Warrick
4A Washington
5A Wayne
5A Wells
5A White
5A Whitley

IOWA

5A Adair
5A Adams
6A Allamakee
5A Appanoose
5A Audubon
5A Benton
6A Black Hawk
5A Boone
6A Bremer
6A Buchanan
6A Buena Vista
6A Butler
6A Calhoun
5A Carroll
5A Cass
5A Cedar
6A Cerro Gordo
6A Cherokee
6A Chickasaw
5A Clarke
6A Clay
6A Clayton
5A Clinton
5A Crawford
5A Dallas
5A Davis
5A Decatur
6A Delaware
5A Des Moines
6A Dickinson
5A Dubuque
6A Emmet
6A Fayette
6A Floyd
6A Franklin
5A Fremont
5A Greene
6A Grundy
5A Guthrie
6A Hamilton
6A Hancock
6A Hardin
5A Harrison
5A Henry
6A Howard
6A Humboldt
6A Ida
5A Iowa
5A Jackson
5A Jasper
5A Jefferson
5A Johnson
5A Jones
5A Keokuk
6A Kossuth
5A Lee
5A Linn
5A Louisa
5A Lucas
6A Lyon
5A Madison
5A Mahaska
5A Marion
5A Marshall
5A Mills
6A Mitchell
5A Monona
5A Monroe
5A Montgomery
5A Muscatine
6A O'Brien
6A Osceola
5A Page
6A Palo Alto
6A Plymouth
6A Pocahontas
5A Polk
5A Pottawattamie
5A Poweshiek
5A Ringgold
6A Sac
5A Scott
5A Shelby
6A Sioux
5A Story
5A Tama
5A Taylor
5A Union
5A Van Buren
5A Wapello
5A Warren
5A Washington
5A Wayne
6A Webster
6A Winnebago

(continued)

TABLE R301.1—continued
CLIMATE ZONES, MOISTURE REGIMES, AND WARM-HUMID DESIGNATIONS BY STATE, COUNTY AND TERRITORY

6A Winneshiek
5A Woodbury
6A Worth
6A Wright

KANSAS

4A Allen
4A Anderson
4A Atchison
4A Barber
4A Barton
4A Bourbon
4A Brown
4A Butler
4A Chase
4A Chautauqua
4A Cherokee
5A Cheyenne
4A Clark
4A Clay
5A Cloud
4A Coffey
4A Comanche
4A Cowley
4A Crawford
5A Decatur
4A Dickinson
4A Doniphan
4A Douglas
4A Edwards
4A Elk
5A Ellis
4A Ellsworth
4A Finney
4A Ford
4A Franklin
4A Geary
5A Gove
5A Graham
4A Grant
4A Gray
5A Greeley
4A Greenwood
5A Hamilton
4A Harper
4A Harvey
4A Haskell
4A Hodgeman
4A Jackson
4A Jefferson
5A Jewell
4A Johnson
4A Kearny
4A Kingman
4A Kiowa
4A Labette
5A Lane
4A Leavenworth
4A Lincoln
4A Linn
5A Logan
4A Lyon
4A Marion
4A Marshall
4A McPherson
4A Meade
4A Miami
5A Mitchell
4A Montgomery
4A Morris
4A Morton
4A Nemaha
4A Neosho
5A Ness
5A Norton
4A Osage
5A Osborne
4A Ottawa
4A Pawnee
5A Phillips
4A Pottawatomie
4A Pratt
5A Rawlins
4A Reno
5A Republic
4A Rice
4A Riley
5A Rooks
4A Rush
4A Russell
4A Saline
5A Scott
4A Sedgwick
4A Seward
4A Shawnee
5A Sheridan
5A Sherman
5A Smith
4A Stafford
4A Stanton
4A Stevens
4A Sumner
5A Thomas
5A Trego
4A Wabaunsee
5A Wallace
4A Washington
5A Wichita
4A Wilson
4A Woodson
4A Wyandotte

KENTUCKY

4A (all)

LOUISIANA

2A Acadia*
2A Allen*
2A Ascension*
2A Assumption*
2A Avoyelles*
2A Beauregard*
3A Bienville*
3A Bossier*
3A Caddo*
2A Calcasieu*
3A Caldwell*
2A Cameron*
3A Catahoula*
3A Claiborne*
3A Concordia*
3A De Soto*
2A East Baton Rouge*
3A East Carroll
2A East Feliciana*
2A Evangeline*
3A Franklin*
3A Grant*
2A Iberia*
2A Iberville*
3A Jackson*
2A Jefferson*
2A Jefferson Davis*
2A Lafayette*
2A Lafourche*
3A La Salle*
3A Lincoln*
2A Livingston*
3A Madison*
3A Morehouse
3A Natchitoches*
2A Orleans*
3A Ouachita*
2A Plaquemines*
2A Pointe Coupee*
2A Rapides*
3A Red River*
3A Richland*
3A Sabine*
2A St. Bernard*
2A St. Charles*
2A St. Helena*
2A St. James*
2A St. John the Baptist*
2A St. Landry*
2A St. Martin*
2A St. Mary*
2A St. Tammany*
2A Tangipahoa*
3A Tensas*
2A Terrebonne*
3A Union*
2A Vermilion*
3A Vernon*
2A Washington*
3A Webster*
2A West Baton Rouge*
3A West Carroll
2A West Feliciana*
3A Winn*

MAINE

6A Androscoggin
7 Aroostook
6A Cumberland
6A Franklin
6A Hancock
6A Kennebec
6A Knox
6A Lincoln
6A Oxford
6A Penobscot
6A Piscataquis
6A Sagadahoc
6A Somerset
6A Waldo
6A Washington
6A York

MARYLAND

4A Allegany
4A Anne Arundel
4A Baltimore
4A Baltimore (city)
4A Calvert
4A Caroline
4A Carroll
4A Cecil
4A Charles
4A Dorchester
4A Frederick
5A Garrett
4A Harford
4A Howard
4A Kent
4A Montgomery
4A Prince George's
4A Queen Anne's
4A Somerset
4A St. Mary's
4A Talbot
4A Washington
4A Wicomico
4A Worcester

MASSACHSETTS

5A (all)

MICHIGAN

6A Alcona
6A Alger

(continued)

TABLE R301.1—continued
CLIMATE ZONES, MOISTURE REGIMES, AND WARM-HUMID DESIGNATIONS BY STATE, COUNTY AND TERRITORY

5A Allegan
6A Alpena
6A Antrim
6A Arenac
7 Baraga
5A Barry
5A Bay
6A Benzie
5A Berrien
5A Branch
5A Calhoun
5A Cass
6A Charlevoix
6A Cheboygan
7 Chippewa
6A Clare
5A Clinton
6A Crawford
6A Delta
6A Dickinson
5A Eaton
6A Emmet
5A Genesee
6A Gladwin
7 Gogebic
6A Grand Traverse
5A Gratiot
5A Hillsdale
7 Houghton
6A Huron
5A Ingham
5A Ionia
6A Iosco
7 Iron
6A Isabella
5A Jackson
5A Kalamazoo
6A Kalkaska
5A Kent
7 Keweenaw
6A Lake
5A Lapeer
6A Leelanau
5A Lenawee
5A Livingston
7 Luce
7 Mackinac
5A Macomb
6A Manistee
6A Marquette
6A Mason
6A Mecosta
6A Menominee
5A Midland
6A Missaukee
5A Monroe
5A Montcalm
6A Montmorency
5A Muskegon
6A Newaygo
5A Oakland
6A Oceana
6A Ogemaw
7 Ontonagon
6A Osceola
6A Oscoda
6A Otsego
5A Ottawa
6A Presque Isle
6A Roscommon
5A Saginaw
6A Sanilac
7 Schoolcraft
5A Shiawassee
5A St. Clair
5A St. Joseph
5A Tuscola
5A Van Buren
5A Washtenaw
5A Wayne
6A Wexford

MINNESOTA

7 Aitkin
6A Anoka
7 Becker
7 Beltrami
6A Benton
6A Big Stone
6A Blue Earth
6A Brown
7 Carlton
6A Carver
7 Cass
6A Chippewa
6A Chisago
7 Clay
7 Clearwater
7 Cook
6A Cottonwood
7 Crow Wing
6A Dakota
6A Dodge
6A Douglas
6A Faribault
6A Fillmore
6A Freeborn
6A Goodhue
7 Grant
6A Hennepin
6A Houston
7 Hubbard
6A Isanti
7 Itasca
6A Jackson
7 Kanabec
6A Kandiyohi
7 Kittson
7 Koochiching
6A Lac qui Parle
7 Lake
7 Lake of the Woods
6A Le Sueur
6A Lincoln
6A Lyon
7 Mahnomen
7 Marshall
6A Martin
6A McLeod
6A Meeker
7 Mille Lacs
6A Morrison
6A Mower
6A Murray
6A Nicollet
6A Nobles
7 Norman
6A Olmsted
7 Otter Tail
7 Pennington
7 Pine
6A Pipestone
7 Polk
6A Pope
6A Ramsey
7 Red Lake
6A Redwood
6A Renville
6A Rice
6A Rock
7 Roseau
6A Scott
6A Sherburne
6A Sibley
6A Stearns
6A Steele
6A Stevens
7 St. Louis
6A Swift
6A Todd
6A Traverse
6A Wabasha
7 Wadena
6A Waseca
6A Washington
6A Watonwan
7 Wilkin
6A Winona
6A Wright
6A Yellow Medicine

MISSISSIPPI

3A Adams*
3A Alcorn
3A Amite*
3A Attala
3A Benton
3A Bolivar
3A Calhoun
3A Carroll
3A Chickasaw
3A Choctaw
3A Claiborne*
3A Clarke
3A Clay
3A Coahoma
3A Copiah*
3A Covington*
3A DeSoto
3A Forrest*
3A Franklin*
3A George*
3A Greene*
3A Grenada
2A Hancock*
2A Harrison*
3A Hinds*
3A Holmes
3A Humphreys
3A Issaquena
3A Itawamba
2A Jackson*
3A Jasper
3A Jefferson*
3A Jefferson Davis*
3A Jones*
3A Kemper
3A Lafayette
3A Lamar*
3A Lauderdale
3A Lawrence*
3A Leake
3A Lee
3A Leflore
3A Lincoln*
3A Lowndes
3A Madison
3A Marion*
3A Marshall
3A Monroe
3A Montgomery
3A Neshoba
3A Newton
3A Noxubee
3A Oktibbeha
3A Panola
2A Pearl River*
3A Perry*
3A Pike*

(continued)

**TABLE R301.1—continued
CLIMATE ZONES, MOISTURE REGIMES, AND WARM-HUMID DESIGNATIONS BY STATE, COUNTY AND TERRITORY**

3A Pontotoc
3A Prentiss
3A Quitman
3A Rankin*
3A Scott
3A Sharkey
3A Simpson*
3A Smith*
2A Stone*
3A Sunflower
3A Tallahatchie
3A Tate
3A Tippah
3A Tishomingo
3A Tunica
3A Union
3A Walthall*
3A Warren*
3A Washington
3A Wayne*
3A Webster
3A Wilkinson*
3A Winston
3A Yalobusha
3A Yazoo

MISSOURI

5A Adair
5A Andrew
5A Atchison
4A Audrain
4A Barry
4A Barton
4A Bates
4A Benton
4A Bollinger
4A Boone
5A Buchanan
4A Butler
5A Caldwell
4A Callaway
4A Camden
4A Cape Girardeau
4A Carroll
4A Carter
4A Cass
4A Cedar
5A Chariton
4A Christian
5A Clark
4A Clay
5A Clinton
4A Cole
4A Cooper
4A Crawford
4A Dade
4A Dallas
5A Daviess
5A DeKalb
4A Dent
4A Douglas
4A Dunklin
4A Franklin
4A Gasconade
5A Gentry
4A Greene
5A Grundy
5A Harrison
4A Henry
4A Hickory
5A Holt
4A Howard
4A Howell
4A Iron
4A Jackson
4A Jasper
4A Jefferson
4A Johnson
5A Knox
4A Laclede
4A Lafayette
4A Lawrence
5A Lewis
4A Lincoln
5A Linn
5A Livingston
5A Macon
4A Madison
4A Maries
5A Marion
4A McDonald
5A Mercer
4A Miller
4A Mississippi
4A Moniteau
4A Monroe
4A Montgomery
4A Morgan
4A New Madrid
4A Newton
5A Nodaway
4A Oregon
4A Osage
4A Ozark
4A Pemiscot
4A Perry
4A Pettis
4A Phelps
5A Pike
4A Platte
4A Polk
4A Pulaski
5A Putnam
5A Ralls
4A Randolph
4A Ray
4A Reynolds
4A Ripley
4A Saline
5A Schuyler
5A Scotland
4A Scott
4A Shannon
5A Shelby
4A St. Charles
4A St. Clair
4A St. Francois
4A St. Louis
4A St. Louis (city)
4A Ste. Genevieve
4A Stoddard
4A Stone
5A Sullivan
4A Taney
4A Texas
4A Vernon
4A Warren
4A Washington
4A Wayne
4A Webster
5A Worth
4A Wright

MONTANA

6B (all)

NEBRASKA

5A (all)

NEVADA

5B Carson City (city)
5B Churchill
3B Clark
5B Douglas
5B Elko
5B Esmeralda
5B Eureka
5B Humboldt
5B Lander
5B Lincoln
5B Lyon
5B Mineral
5B Nye
5B Pershing
5B Storey
5B Washoe
5B White Pine

NEW HAMPSHIRE

6A Belknap
6A Carroll
5A Cheshire
6A Coos
6A Grafton
5A Hillsborough
6A Merrimack
5A Rockingham
5A Strafford
6A Sullivan

NEW JERSEY

4A Atlantic
5A Bergen
4A Burlington
4A Camden
4A Cape May
4A Cumberland
4A Essex
4A Gloucester
4A Hudson
5A Hunterdon
5A Mercer
4A Middlesex
4A Monmouth
5A Morris
4A Ocean
5A Passaic
4A Salem
5A Somerset
5A Sussex
4A Union
5A Warren

NEW MEXICO

4B Bernalillo
5B Catron
3B Chaves
4B Cibola
5B Colfax
4B Curry
4B DeBaca
3B Dona Ana
3B Eddy
4B Grant
4B Guadalupe
5B Harding
3B Hidalgo
3B Lea
4B Lincoln
5B Los Alamos
3B Luna
5B McKinley
5B Mora
3B Otero
4B Quay
5B Rio Arriba
4B Roosevelt
5B Sandoval
5B San Juan
5B San Miguel
5B Santa Fe
4B Sierra
4B Socorro

(continued)

TABLE R301.1—continued
CLIMATE ZONES, MOISTURE REGIMES, AND WARM-HUMID DESIGNATIONS BY STATE, COUNTY AND TERRITORY

5B Taos
5B Torrance
4B Union
4B Valencia

NEW YORK

5A Albany
6A Allegany
4A Bronx
6A Broome
6A Cattaraugus
5A Cayuga
5A Chautauqua
5A Chemung
6A Chenango
6A Clinton
5A Columbia
5A Cortland
6A Delaware
5A Dutchess
5A Erie
6A Essex
6A Franklin
6A Fulton
5A Genesee
5A Greene
6A Hamilton
6A Herkimer
6A Jefferson
4A Kings
6A Lewis
5A Livingston
6A Madison
5A Monroe
6A Montgomery
4A Nassau
4A New York
5A Niagara
6A Oneida
5A Onondaga
5A Ontario
5A Orange
5A Orleans
5A Oswego
6A Otsego
5A Putnam
4A Queens
5A Rensselaer
4A Richmond
5A Rockland
5A Saratoga
5A Schenectady
6A Schoharie
6A Schuyler
5A Seneca
6A Steuben
6A St. Lawrence
4A Suffolk
6A Sullivan
5A Tioga
6A Tompkins
6A Ulster
6A Warren
5A Washington
5A Wayne
4A Westchester
6A Wyoming
5A Yates

NORTH CAROLINA

4A Alamance
4A Alexander
5A Alleghany
3A Anson
5A Ashe
5A Avery
3A Beaufort
4A Bertie
3A Bladen
3A Brunswick*
4A Buncombe
4A Burke
3A Cabarrus
4A Caldwell
3A Camden
3A Carteret*
4A Caswell
4A Catawba
4A Chatham
4A Cherokee
3A Chowan
4A Clay
4A Cleveland
3A Columbus*
3A Craven
3A Cumberland
3A Currituck
3A Dare
3A Davidson
4A Davie
3A Duplin
4A Durham
3A Edgecombe
4A Forsyth
4A Franklin
3A Gaston
4A Gates
4A Graham
4A Granville
3A Greene
4A Guilford
4A Halifax
4A Harnett
4A Haywood
4A Henderson
4A Hertford
3A Hoke
3A Hyde
4A Iredell
4A Jackson
3A Johnston
3A Jones
4A Lee
3A Lenoir
4A Lincoln
4A Macon
4A Madison
3A Martin
4A McDowell
3A Mecklenburg
5A Mitchell
3A Montgomery
3A Moore
4A Nash
3A New Hanover*
4A Northampton
3A Onslow*
4A Orange
3A Pamlico
3A Pasquotank
3A Pender*
3A Perquimans
4A Person
3A Pitt
4A Polk
3A Randolph
3A Richmond
3A Robeson
4A Rockingham
3A Rowan
4A Rutherford
3A Sampson
3A Scotland
3A Stanly
4A Stokes
4A Surry
4A Swain
4A Transylvania
3A Tyrrell
3A Union
4A Vance
4A Wake
4A Warren
3A Washington
5A Watauga
3A Wayne
4A Wilkes
3A Wilson
4A Yadkin
5A Yancey

NORTH DAKOTA

6A Adams
7 Barnes
7 Benson
6A Billings
7 Bottineau
6A Bowman
7 Burke
6A Burleigh
7 Cass
7 Cavalier
6A Dickey
7 Divide
6A Dunn
7 Eddy
6A Emmons
7 Foster
6A Golden Valley
7 Grand Forks
6A Grant
7 Griggs
6A Hettinger
7 Kidder
6A LaMoure
6A Logan
7 McHenry
6A McIntosh
6A McKenzie
7 McLean
6A Mercer
6A Morton
7 Mountrail
7 Nelson
6A Oliver
7 Pembina
7 Pierce
7 Ramsey
6A Ransom
7 Renville
6A Richland
7 Rolette
6A Sargent
7 Sheridan
6A Sioux
6A Slope
6A Stark
7 Steele
7 Stutsman
7 Towner
7 Traill
7 Walsh
7 Ward
7 Wells
7 Williams

OHIO

4A Adams
5A Allen

(continued)

TABLE R301.1—continued
CLIMATE ZONES, MOISTURE REGIMES, AND WARM-HUMID DESIGNATIONS BY STATE, COUNTY AND TERRITORY

5A Ashland
5A Ashtabula
5A Athens
5A Auglaize
5A Belmont
4A Brown
5A Butler
5A Carroll
5A Champaign
5A Clark
4A Clermont
5A Clinton
5A Columbiana
5A Coshocton
5A Crawford
5A Cuyahoga
5A Darke
5A Defiance
5A Delaware
5A Erie
5A Fairfield
5A Fayette
5A Franklin
5A Fulton
4A Gallia
5A Geauga
5A Greene
5A Guernsey
4A Hamilton
5A Hancock
5A Hardin
5A Harrison
5A Henry
5A Highland
5A Hocking
5A Holmes
5A Huron
5A Jackson
5A Jefferson
5A Knox
5A Lake
4A Lawrence
5A Licking
5A Logan
5A Lorain
5A Lucas
5A Madison
5A Mahoning
5A Marion
5A Medina
5A Meigs
5A Mercer
5A Miami
5A Monroe
5A Montgomery
5A Morgan
5A Morrow
5A Muskingum
5A Noble
5A Ottawa
5A Paulding
5A Perry
5A Pickaway
4A Pike
5A Portage
5A Preble
5A Putnam
5A Richland
5A Ross
5A Sandusky
4A Scioto
5A Seneca
5A Shelby
5A Stark
5A Summit
5A Trumbull
5A Tuscarawas
5A Union
5A Van Wert
5A Vinton
5A Warren
4A Washington
5A Wayne
5A Williams
5A Wood
5A Wyandot

OKLAHOMA

3A Adair
3A Alfalfa
3A Atoka
4B Beaver
3A Beckham
3A Blaine
3A Bryan
3A Caddo
3A Canadian
3A Carter
3A Cherokee
3A Choctaw
4B Cimarron
3A Cleveland
3A Coal
3A Comanche
3A Cotton
3A Craig
3A Creek
3A Custer
3A Delaware
3A Dewey
3A Ellis
3A Garfield
3A Garvin
3A Grady
3A Grant
3A Greer
3A Harmon
3A Harper
3A Haskell
3A Hughes
3A Jackson
3A Jefferson
3A Johnston
3A Kay
3A Kingfisher
3A Kiowa
3A Latimer
3A Le Flore
3A Lincoln
3A Logan
3A Love
3A Major
3A Marshall
3A Mayes
3A McClain
3A McCurtain
3A McIntosh
3A Murray
3A Muskogee
3A Noble
3A Nowata
3A Okfuskee
3A Oklahoma
3A Okmulgee
3A Osage
3A Ottawa
3A Pawnee
3A Payne
3A Pittsburg
3A Pontotoc
3A Pottawatomie
3A Pushmataha
3A Roger Mills
3A Rogers
3A Seminole
3A Sequoyah
3A Stephens
4B Texas
3A Tillman
3A Tulsa
3A Wagoner
3A Washington
3A Washita
3A Woods
3A Woodward

OREGON

5B Baker
4C Benton
4C Clackamas
4C Clatsop
4C Columbia
4C Coos
5B Crook
4C Curry
5B Deschutes
4C Douglas
5B Gilliam
5B Grant
5B Harney
5B Hood River
4C Jackson
5B Jefferson
4C Josephine
5B Klamath
5B Lake
4C Lane
4C Lincoln
4C Linn
5B Malheur
4C Marion
5B Morrow
4C Multnomah
4C Polk
5B Sherman
4C Tillamook
5B Umatilla
5B Union
5B Wallowa
5B Wasco
4C Washington
5B Wheeler
4C Yamhill

PENNSYLVANIA

5A Adams
5A Allegheny
5A Armstrong
5A Beaver
5A Bedford
5A Berks
5A Blair
5A Bradford
4A Bucks
5A Butler
5A Cambria
6A Cameron
5A Carbon
5A Centre
4A Chester
5A Clarion
6A Clearfield
5A Clinton
5A Columbia
5A Crawford
5A Cumberland
5A Dauphin
4A Delaware
6A Elk
5A Erie
5A Fayette
5A Forest
5A Franklin
5A Fulton
5A Greene

(continued)

TABLE R301.1—continued
CLIMATE ZONES, MOISTURE REGIMES, AND WARM-HUMID DESIGNATIONS BY STATE, COUNTY AND TERRITORY

5A Huntingdon
5A Indiana
5A Jefferson
5A Juniata
5A Lackawanna
5A Lancaster
5A Lawrence
5A Lebanon
5A Lehigh
5A Luzerne
5A Lycoming
6A McKean
5A Mercer
5A Mifflin
5A Monroe
4A Montgomery
5A Montour
5A Northampton
5A Northumberland
5A Perry
4A Philadelphia
5A Pike
6A Potter
5A Schuylkill
5A Snyder
5A Somerset
5A Sullivan
6A Susquehanna
6A Tioga
5A Union
5A Venango
5A Warren
5A Washington
6A Wayne
5A Westmoreland
5A Wyoming
4A York

RHODE ISLAND

5A (all)

SOUTH CAROLINA

3A Abbeville
3A Aiken
3A Allendale*
3A Anderson
3A Bamberg*
3A Barnwell*
3A Beaufort*
3A Berkeley*
3A Calhoun
3A Charleston*
3A Cherokee
3A Chester
3A Chesterfield
3A Clarendon
3A Colleton*
3A Darlington
3A Dillon
3A Dorchester*
3A Edgefield
3A Fairfield
3A Florence
3A Georgetown*
3A Greenville
3A Greenwood
3A Hampton*
3A Horry*
3A Jasper*
3A Kershaw
3A Lancaster
3A Laurens
3A Lee
3A Lexington
3A Marion
3A Marlboro
3A McCormick
3A Newberry
3A Oconee
3A Orangeburg
3A Pickens
3A Richland
3A Saluda
3A Spartanburg
3A Sumter
3A Union
3A Williamsburg
3A York

SOUTH DAKOTA

6A Aurora
6A Beadle
5A Bennett
5A Bon Homme
6A Brookings
6A Brown
6A Brule
6A Buffalo
6A Butte
6A Campbell
5A Charles Mix
6A Clark
5A Clay
6A Codington
6A Corson
6A Custer
6A Davison
6A Day
6A Deuel
6A Dewey
5A Douglas
6A Edmunds
6A Fall River
6A Faulk
6A Grant
5A Gregory
6A Haakon
6A Hamlin
6A Hand
6A Hanson
6A Harding
6A Hughes
5A Hutchinson
6A Hyde
5A Jackson
6A Jerauld
6A Jones
6A Kingsbury
6A Lake
6A Lawrence
6A Lincoln
6A Lyman
6A Marshall
6A McCook
6A McPherson
6A Meade
5A Mellette
6A Miner
6A Minnehaha
6A Moody
6A Pennington
6A Perkins
6A Potter
6A Roberts
6A Sanborn
6A Shannon
6A Spink
6A Stanley
6A Sully
5A Todd
5A Tripp
6A Turner
5A Union
6A Walworth
5A Yankton
6A Ziebach

TENNESSEE

4A Anderson
4A Bedford
4A Benton
4A Bledsoe
4A Blount
4A Bradley
4A Campbell
4A Cannon
4A Carroll
4A Carter
4A Cheatham
3A Chester
4A Claiborne
4A Clay
4A Cocke
4A Coffee
3A Crockett
4A Cumberland
4A Davidson
4A Decatur
4A DeKalb
4A Dickson
3A Dyer
3A Fayette
4A Fentress
4A Franklin
4A Gibson
4A Giles
4A Grainger
4A Greene
4A Grundy
4A Hamblen
4A Hamilton
4A Hancock
3A Hardeman
3A Hardin
4A Hawkins
3A Haywood
3A Henderson
4A Henry
4A Hickman
4A Houston
4A Humphreys
4A Jackson
4A Jefferson
4A Johnson
4A Knox
3A Lake
3A Lauderdale
4A Lawrence
4A Lewis
4A Lincoln
4A Loudon
4A Macon
3A Madison
4A Marion
4A Marshall
4A Maury
4A McMinn
3A McNairy
4A Meigs
4A Monroe
4A Montgomery
4A Moore
4A Morgan
4A Obion
4A Overton
4A Perry
4A Pickett
4A Polk
4A Putnam
4A Rhea

(continued)

TABLE R301.1—continued
CLIMATE ZONES, MOISTURE REGIMES, AND WARM-HUMID DESIGNATIONS BY STATE, COUNTY AND TERRITORY

4A Roane
4A Robertson
4A Rutherford
4A Scott
4A Sequatchie
4A Sevier
3A Shelby
4A Smith
4A Stewart
4A Sullivan
4A Sumner
3A Tipton
4A Trousdale
4A Unicoi
4A Union
4A Van Buren
4A Warren
4A Washington
4A Wayne
4A Weakley
4A White
4A Williamson
4A Wilson

TEXAS

2A Anderson*
3B Andrews
2A Angelina*
2A Aransas*
3A Archer
4B Armstrong
2A Atascosa*
2A Austin*
4B Bailey
2B Bandera
2A Bastrop*
3B Baylor
2A Bee*
2A Bell*
2A Bexar*
3A Blanco*
3B Borden
2A Bosque*
3A Bowie*
2A Brazoria*
2A Brazos*
3B Brewster
4B Briscoe
2A Brooks*
3A Brown*
2A Burleson*
3A Burnet*
2A Caldwell*
2A Calhoun*
3B Callahan
2A Cameron*
3A Camp*
4B Carson
3A Cass*
4B Castro
2A Chambers*
2A Cherokee*
3B Childress
3A Clay
4B Cochran
3B Coke
3B Coleman
3A Collin*
3B Collingsworth
2A Colorado*
2A Comal*
3A Comanche*
3B Concho
3A Cooke
2A Coryell*
3B Cottle
3B Crane
3B Crockett
3B Crosby
3B Culberson
4B Dallam
3A Dallas*
3B Dawson
4B Deaf Smith
3A Delta
3A Denton*
2A DeWitt*
3B Dickens
2B Dimmit
4B Donley
2A Duval*
3A Eastland
3B Ector
2B Edwards
3A Ellis*
3B El Paso
3A Erath*
2A Falls*
3A Fannin
2A Fayette*
3B Fisher
4B Floyd
3B Foard
2A Fort Bend*
3A Franklin*
2A Freestone*
2B Frio
3B Gaines
2A Galveston*
3B Garza
3A Gillespie*
3B Glasscock
2A Goliad*
2A Gonzales*
4B Gray
3A Grayson
3A Gregg*
2A Grimes*
2A Guadalupe*
4B Hale
3B Hall
3A Hamilton*
4B Hansford
3B Hardeman
2A Hardin*
2A Harris*
3A Harrison*
4B Hartley
3B Haskell
2A Hays*
3B Hemphill
3A Henderson*
2A Hidalgo*
2A Hill*
4B Hockley
3A Hood*
3A Hopkins*
2A Houston*
3B Howard
3B Hudspeth
3A Hunt*
4B Hutchinson
3B Irion
3A Jack
2A Jackson*
2A Jasper*
3B Jeff Davis
2A Jefferson*
2A Jim Hogg*
2A Jim Wells*
3A Johnson*
3B Jones
2A Karnes*
3A Kaufman*
3A Kendall*
2A Kenedy*
3B Kent
3B Kerr
3B Kimble
3B King
2B Kinney
2A Kleberg*
3B Knox
3A Lamar*
4B Lamb
3A Lampasas*
2B La Salle
2A Lavaca*
2A Lee*
2A Leon*
2A Liberty*
2A Limestone*
4B Lipscomb
2A Live Oak*
3A Llano*
3B Loving
3B Lubbock
3B Lynn
2A Madison*
3A Marion*
3B Martin
3B Mason
2A Matagorda*
2B Maverick
3B McCulloch
2A McLennan*
2A McMullen*
2B Medina
3B Menard
3B Midland
2A Milam*
3A Mills*
3B Mitchell
3A Montague
2A Montgomery*
4B Moore
3A Morris*
3B Motley
3A Nacogdoches*
3A Navarro*
2A Newton*
3B Nolan
2A Nueces*
4B Ochiltree
4B Oldham
2A Orange*
3A Palo Pinto*
3A Panola*
3A Parker*
4B Parmer
3B Pecos
2A Polk*
4B Potter
3B Presidio
3A Rains*
4B Randall
3B Reagan
2B Real
3A Red River*
3B Reeves
2A Refugio*
4B Roberts
2A Robertson*
3A Rockwall*
3B Runnels
3A Rusk*
3A Sabine*
3A San Augustine*
2A San Jacinto*
2A San Patricio*

(continued)

TABLE R301.1—continued
CLIMATE ZONES, MOISTURE REGIMES, AND WARM-HUMID DESIGNATIONS BY STATE, COUNTY AND TERRITORY

3A San Saba*
3B Schleicher
3B Scurry
3B Shackelford
3A Shelby*
4B Sherman
3A Smith*
3A Somervell*
2A Starr*
3A Stephens
3B Sterling
3B Stonewall
3B Sutton
4B Swisher
3A Tarrant*
3B Taylor
3B Terrell
3B Terry
3B Throckmorton
3A Titus*
3B Tom Green
2A Travis*
2A Trinity*
2A Tyler*
3A Upshur*
3B Upton
2B Uvalde
2B Val Verde
3A Van Zandt*
2A Victoria*
2A Walker*
2A Waller*
3B Ward
2A Washington*
2B Webb
2A Wharton*
3B Wheeler
3A Wichita
3B Wilbarger
2A Willacy*
2A Williamson*
2A Wilson*
3B Winkler
3A Wise
3A Wood*
4B Yoakum
3A Young
2B Zapata
2B Zavala

UTAH

5B Beaver
6B Box Elder
6B Cache
6B Carbon
6B Daggett
5B Davis
6B Duchesne
5B Emery
5B Garfield
5B Grand
5B Iron
5B Juab
5B Kane
5B Millard
6B Morgan
5B Piute
6B Rich
5B Salt Lake
5B San Juan
5B Sanpete
5B Sevier
6B Summit
5B Tooele
6B Uintah
5B Utah
6B Wasatch
3B Washington
5B Wayne
5B Weber

VERMONT

6A (all)

VIRGINIA

4A (all)

WASHINGTON

5B Adams
5B Asotin
5B Benton
5B Chelan
4C Clallam
4C Clark
5B Columbia
4C Cowlitz
5B Douglas
6B Ferry
5B Franklin
5B Garfield
5B Grant
4C Grays Harbor
4C Island
4C Jefferson
4C King
4C Kitsap
5B Kittitas
5B Klickitat
4C Lewis
5B Lincoln
4C Mason
6B Okanogan
4C Pacific
6B Pend Oreille
4C Pierce
4C San Juan
4C Skagit
5B Skamania
4C Snohomish
5B Spokane
6B Stevens
4C Thurston
4C Wahkiakum
5B Walla Walla
4C Whatcom
5B Whitman
5B Yakima

WEST VIRGINIA

5A Barbour
4A Berkeley
4A Boone
4A Braxton
5A Brooke
4A Cabell
4A Calhoun
4A Clay
5A Doddridge
5A Fayette
4A Gilmer
5A Grant
5A Greenbrier
5A Hampshire
5A Hancock
5A Hardy
5A Harrison
4A Jackson
4A Jefferson
4A Kanawha
5A Lewis
4A Lincoln
4A Logan
5A Marion
5A Marshall
4A Mason
4A McDowell
4A Mercer
5A Mineral
4A Mingo
5A Monongalia
4A Monroe
4A Morgan
5A Nicholas
5A Ohio
5A Pendleton
4A Pleasants
5A Pocahontas
5A Preston
4A Putnam
5A Raleigh
5A Randolph
4A Ritchie
4A Roane
5A Summers
5A Taylor
5A Tucker
4A Tyler
5A Upshur
4A Wayne
5A Webster
5A Wetzel
4A Wirt
4A Wood
4A Wyoming

WISCONSIN

6A Adams
7 Ashland
6A Barron
7 Bayfield
6A Brown
6A Buffalo
7 Burnett
6A Calumet
6A Chippewa
6A Clark
6A Columbia
6A Crawford
6A Dane
6A Dodge
6A Door
7 Douglas
6A Dunn
6A Eau Claire
7 Florence
6A Fond du Lac
7 Forest
6A Grant
6A Green
6A Green Lake
6A Iowa
7 Iron
6A Jackson
6A Jefferson
6A Juneau
6A Kenosha
6A Kewaunee
6A La Crosse
6A Lafayette
7 Langlade
7 Lincoln
6A Manitowoc
6A Marathon
6A Marinette
6A Marquette
6A Menominee
6A Milwaukee
6A Monroe
6A Oconto
7 Oneida
6A Outagamie

(continued)

TABLE R301.1—continued
CLIMATE ZONES, MOISTURE REGIMES, AND WARM-HUMID DESIGNATIONS BY STATE, COUNTY AND TERRITORY

6A Ozaukee
6A Pepin
6A Pierce
6A Polk
6A Portage
7 Price
6A Racine
6A Richland
6A Rock
6A Rusk
6A Sauk
7 Sawyer
6A Shawano
6A Sheboygan
6A St. Croix
7 Taylor
6A Trempealeau
6A Vernon
7 Vilas
6A Walworth
7 Washburn
6A Washington
6A Waukesha
6A Waupaca
6A Waushara
6A Winnebago
6A Wood

WYOMING

6B Albany
6B Big Horn
6B Campbell
6B Carbon
6B Converse
6B Crook
6B Fremont
5B Goshen
6B Hot Springs
6B Johnson
6B Laramie
7 Lincoln
6B Natrona
6B Niobrara
6B Park
5B Platte
6B Sheridan
7 Sublette
6B Sweetwater
7 Teton
6B Uinta
6B Washakie
6B Weston

US TERRITORIES

AMERICAN SAMOA

1A (all)*

GUAM

1A (all)*

NORTHERN MARIANA ISLANDS

1A (all)*

PUERTO RICO

1A (all)*

VIRGIN ISLANDS

1A (all)*

TABLE R301.3(1)
INTERNATIONAL CLIMATE ZONE DEFINITIONS

MAJOR CLIMATE-TYPE DEFINITIONS
Marine (C) Definition—Locations meeting all four criteria: 1. Mean temperature of coldest month between -3°C (27°F) and 18°C (65°F). 2. Warmest month mean < 22°C (72°F). 3. Not fewer than four months with mean temperatures over 10°C (50°F). 4. Dry season in summer. The month with the heaviest precipitation in the cold season has not less than three times as much precipitation as the month with the least precipitation in the rest of the year. The cold season is October through March in the Northern Hemisphere and April through September in the Southern Hemisphere.
Dry (B) Definition—Locations meeting the following criteria: Not marine and $P_{in} < 0.44 \times (TF - 19.5)$ [$P_{cm} < 2.0 \times (TC + 7)$ in SI units] where: P_{in} = Annual precipitation in inches (cm) T = Annual mean temperature in °F (°C)
Moist (A) Definition—Locations that are not marine and not dry.
Warm-humid Definition—Moist (A) locations where either of the following wet-bulb temperature conditions shall occur during the warmest six consecutive months of the year: 1. 67°F (19.4°C) or higher for 3,000 or more hours. 2. 73°F (22.8°C) or higher for 1,500 or more hours.

For SI: °C = [(°F) - 32]/1.8, 1 inch = 2.54 cm.

TABLE R301.3(2)
INTERNATIONAL CLIMATE ZONE DEFINITIONS

ZONE NUMBER	THERMAL CRITERIA	
	IP Units	SI Units
1	9000 < CDD50°F	5000 < CDD10°C
2	6300 < CDD50°F ≤ 9000	3500 < CDD10°C ≤ 5000
3A and 3B	4500 < CDD50°F ≤ 6300 AND HDD65°F ≤ 5400	2500 < CDD10°C ≤ 3500 AND HDD18°C ≤ 3000
4A and 4B	CDD50°F ≤ 4500 AND HDD65°F ≤ 5400	CDD10°C ≤ 2500 AND HDD18°C ≤ 3000
3C	HDD65°F ≤ 3600	HDD18°C ≤ 2000
4C	3600 < HDD65°F ≤ 5400	2000 < HDD18°C ≤ 3000
5	5400 < HDD65°F ≤ 7200	3000 < HDD18°C ≤ 4000
6	7200 < HDD65°F ≤ 9000	4000 < HDD18°C ≤ 5000
7	9000 < HDD65°F ≤ 12600	5000 < HDD18°C ≤ 7000
8	12600 < HDD65°F	7000 < HDD18°C

For SI: °C = [(°F) - 32]/1.8.

SECTION R302
DESIGN CONDITIONS

R302.1 Interior design conditions. The interior design temperatures used for heating and cooling load calculations shall be a maximum of 72°F (22°C) for heating and minimum of 75°F (24°C) for cooling.

SECTION R303
MATERIALS, SYSTEMS AND EQUIPMENT

R303.1 Identification. Materials, systems and equipment shall be identified in a manner that will allow a determination of compliance with the applicable provisions of this code.

R303.1.1 Building thermal envelope insulation. An *R*-value identification mark shall be applied by the manufacturer to each piece of *building thermal envelope* insulation that is 12 inches (305 mm) or greater in width. Alternatively, the insulation installers shall provide a certification that indicates the type, manufacturer and *R*-value of insulation installed in each element of the *building thermal envelope*. For blown-in or sprayed fiberglass and cellulose insulation, the initial installed thickness, settled thickness, settled *R*-value, installed density, coverage area and number of bags installed shall be indicated on the certification. For sprayed polyurethane foam (SPF) insulation, the installed thickness of the areas covered and the *R*-value of the installed thickness shall be indicated on the certification. For insulated siding, the *R*-value shall be on a label on the product's package and shall be indicated on the certification. The insulation installer shall sign, date and post the certification in a conspicuous location on the job site.

Exception: For roof insulation installed above the deck, the *R*-value shall be labeled as required by the material standards specified in Table 1508.2 of the *International Building Code* or Table R906.2 of the *International Residential Code*, as applicable.

R303.1.1.1 Blown-in or sprayed roof and ceiling insulation. The thickness of blown-in or sprayed fiberglass and cellulose roof and ceiling insulation shall be written in inches (mm) on markers that are installed at not less than one for every 300 square feet (28 m^2) throughout the attic space. The markers shall be affixed to the trusses or joists and marked with the minimum initial installed thickness with numbers not less than 1 inch (25 mm) in height. Each marker shall face the attic access opening. The thickness and installed *R*-value of sprayed polyurethane foam insulation shall be indicated on the certification provided by the insulation installer.

R303.1.2 Insulation mark installation. Insulating materials shall be installed such that the manufacturer's *R*-value mark is readily observable at inspection.

R303.1.3 Fenestration product rating. *U*-factors of fenestration products such as windows, doors and *skylights* shall be determined in accordance with NFRC 100.

Exception: Where required, garage door *U*-factors shall be determined in accordance with either NFRC 100 or ANSI/DASMA 105.

U-factors shall be determined by an accredited, independent laboratory, and labeled and certified by the manufacturer.

Products lacking such a labeled *U*-factor shall be assigned a default *U*-factor from Table R303.1.3(1) or R303.1.3(2). The *solar heat gain coefficient* (SHGC) and *visible transmittance* (VT) of glazed fenestration products such as windows, glazed doors and *skylights* shall be determined in accordance with NFRC 200 by an accredited, independent laboratory, and labeled and certified by the manufacturer. Products lacking such a labeled SHGC or VT shall be assigned a default SHGC or VT from Table R303.1.3(3).

R303.1.4 Insulation product rating. The thermal resistance, *R*-value, of insulation shall be determined in accordance with Part 460 of US-FTC CFR Title 16 in units of h • ft^2 • °F/Btu at a mean temperature of 75°F (24°C).

R303.1.4.1 Insulated siding. The thermal resistance, *R*-value, of insulated siding shall be determined in accordance with ASTM C1363. Installation for testing shall be in accordance with the manufacturer's instructions.

TABLE R303.1.3(1)
DEFAULT GLAZED WINDOW, GLASS DOOR AND SKYLIGHT *U*-FACTORS

FRAME TYPE	WINDOW AND GLASS DOOR		SKYLIGHT	
	Single pane	Double pane	Single	Double
Metal	1.20	0.80	2.00	1.30
Metal with Thermal Break	1.10	0.65	1.90	1.10
Nonmetal or Metal Clad	0.95	0.55	1.75	1.05
Glazed Block	0.60			

TABLE R303.1.3(2)
DEFAULT OPAQUE DOOR *U*-FACTORS

DOOR TYPE	OPAQUE *U*-FACTOR
Uninsulated Metal	1.20
Insulated Metal	0.60
Wood	0.50
Insulated, nonmetal edge, not exceeding 45% glazing, any glazing double pane	0.35

TABLE R303.1.3(3)
DEFAULT GLAZED FENESTRATION SHGC AND VT

	SINGLE GLAZED		DOUBLE GLAZED		GLAZED BLOCK
	Clear	Tinted	Clear	Tinted	
SHGC	0.8	0.7	0.7	0.6	0.6
VT	0.6	0.3	0.6	0.3	0.6

R303.2 Installation. Materials, systems and equipment shall be installed in accordance with the manufacturer's instructions and the *International Building Code* or the *International Residential Code*, as applicable.

R303.2.1 Protection of exposed foundation insulation. Insulation applied to the exterior of *basement walls*, crawl space walls and the perimeter of slab-on-grade floors shall have a rigid, opaque and weather-resistant protective covering to prevent the degradation of the insulation's thermal performance. The protective covering shall cover the exposed exterior insulation and extend not less than 6 inches (153 mm) below grade.

R303.3 Maintenance information. Maintenance instructions shall be furnished for equipment and systems that require preventive maintenance. Required regular maintenance actions shall be clearly stated and incorporated on a readily *accessible* label. The label shall include the title or publication number for the operation and maintenance manual for that particular model and type of product.

CHAPTER 4 [RE]

RESIDENTIAL ENERGY EFFICIENCY

User note:

About this chapter: *Chapter 4 provides requirements for the thermal envelope of a building, including minimum insulation values for walls, ceiling and floors; maximum fenestration U-factors; minimum fenestration solar heat gain coefficients; and methods for determining building assembly and a total building U-factor. A performance alternative and an energy rating alternative are also provided to allow for energy code compliance other than by the prescriptive method.*

SECTION R401 GENERAL

R401.1 Scope. This chapter applies to *residential buildings*.

R401.2 Compliance. Projects shall comply with one of the following:

1. Sections R401 through R404.
2. Section R405 and the provisions of Sections R401 through R404 indicated as "Mandatory."
3. The energy rating index (ERI) approach in Section R406.

R401.2.1 Tropical zone. *Residential buildings* in the tropical zone at elevations less than 2,400 feet (731.5 m) above sea level shall be deemed to be in compliance with this chapter provided that the following conditions are met:

1. Not more than one-half of the occupied space is air conditioned.
2. The occupied space is not heated.
3. Solar, wind or other renewable energy source supplies not less than 80 percent of the energy for service water heating.
4. Glazing in *conditioned spaces* has a *solar heat gain coefficient* of less than or equal to 0.40, or has an overhang with a projection factor equal to or greater than 0.30.
5. Permanently installed lighting is in accordance with Section R404.
6. The exterior roof surface complies with one of the options in Table C402.3 or the roof or ceiling has insulation with an *R*-value of R-15 or greater. Where attics are present, attics above the insulation are vented and attics below the insulation are unvented.
7. Roof surfaces have a slope of not less than one-fourth unit vertical in 12 units horizontal (21-percent slope). The finished roof does not have water accumulation areas.
8. Operable fenestration provides a ventilation area of not less than 14 percent of the floor area in each room. Alternatively, equivalent ventilation is provided by a ventilation fan.
9. Bedrooms with *exterior walls* facing two different directions have operable fenestration on *exterior walls* facing two directions.
10. Interior doors to bedrooms are capable of being secured in the open position.
11. A ceiling fan or ceiling fan rough-in is provided for bedrooms and the largest space that is not used as a bedroom.

R401.3 Certificate (Mandatory). A permanent certificate shall be completed by the builder or other *approved* party and posted on a wall in the space where the furnace is located, a utility room or an *approved* location inside the *building*. Where located on an electrical panel, the certificate shall not cover or obstruct the visibility of the circuit directory label, service disconnect label or other required labels. The certificate shall indicate the predominant *R*-values of insulation installed in or on ceilings, roofs, walls, foundation components such as slabs, *basement walls*, crawl space walls and floors and ducts outside *conditioned spaces*; *U*-factors of fenestration and the *solar heat gain coefficient* (SHGC) of fenestration, and the results from any required duct system and *building* envelope air leakage testing performed on the *building*. Where there is more than one value for each component, the certificate shall indicate the value covering the largest area. The certificate shall indicate the types and efficiencies of heating, cooling and service water heating equipment. Where a gas-fired unvented room heater, electric furnace or baseboard electric heater is installed in the residence, the certificate shall indicate "gas-fired unvented room heater," "electric furnace" or "baseboard electric heater," as appropriate. An efficiency shall not be indicated for gas-fired unvented room heaters, electric furnaces and electric baseboard heaters.

SECTION R402 BUILDING THERMAL ENVELOPE

R402.1 General (Prescriptive). The *building thermal envelope* shall comply with the requirements of Sections R402.1.1 through R402.1.5.

Exceptions:

1. The following low-energy *buildings*, or portions thereof, separated from the remainder of the building by *building thermal envelope* assemblies complying with this section shall be exempt from the *building thermal envelope* provisions of Section R402.
   - 1.1. Those with a peak design rate of energy usage less than 3.4 Btu/h • ft^2 (10.7 W/m^2) or 1.0 watt/ft^2 of floor area for space-conditioning purposes.
   - 1.2. Those that do not contain *conditioned space*.
2. Log homes designed in accordance with ICC 400.

R402.1.1 Vapor retarder. Wall assemblies in the *building thermal envelope* shall comply with the vapor retarder requirements of Section R702.7 of the *International Residential Code* or Section 1404.3 of the *International Building Code,* as applicable.

R402.1.2 Insulation and fenestration criteria. The *building thermal envelope* shall meet the requirements of Table R402.1.2, based on the *climate zone* specified in Chapter 3.

R402.1.3 *R*-value computation. Insulation material used in layers, such as framing *cavity insulation* or continuous insulation, shall be summed to compute the corresponding component *R*-value. The manufacturer's settled *R*-value shall be used for blown-in insulation. Computed *R*-values shall not include an *R*-value for other building materials or air films. Where insulated siding is used for the purpose of complying with the continuous insulation requirements of Table R402.1.2, the manufacturer's labeled *R*-value for the insulated siding shall be reduced by R-0.6.

R402.1.4 *U*-factor alternative. An assembly with a *U*-factor equal to or less than that specified in Table R402.1.4 shall be an alternative to the *R*-value in Table R402.1.2.

TABLE R402.1.2
INSULATION AND FENESTRATION REQUIREMENTS BY COMPONENT[a]

CLIMATE ZONE	FENESTRATION *U*-FACTOR[b]	SKYLIGHT[b] *U*-FACTOR	GLAZED FENESTRATION SHGC[b, e]	CEILING *R*-VALUE	WOOD FRAME WALL *R*-VALUE	MASS WALL *R*-VALUE[i]	FLOOR *R*-VALUE	BASEMENT[c] WALL *R*-VALUE	SLAB[d] *R*-VALUE & DEPTH	CRAWL SPACE[c] WALL *R*-VALUE
1	NR	0.75	0.25	30	13	3/4	13	0	0	0
2	0.40	0.65	0.25	38	13	4/6	13	0	0	0
3	0.32	0.55	0.25	38	20 or 13+5[h]	8/13	19	5/13[f]	0	5/13
4 except Marine	0.32	0.55	0.40	49	20 or 13+5[h]	8/13	19	10 /13	10, 2 ft	10/13
5 and Marine 4	0.30	0.55	NR	49	20 or 13+5[h]	13/17	30[g]	15/19	10, 2 ft	15/19
6	0.30	0.55	NR	49	20+5[h] or 13+10[h]	15/20	30[g]	15/19	10, 4 ft	15/19
7 and 8	0.30	0.55	NR	49	20+5[h] or 13+10[h]	19/21	38[g]	15/19	10, 4 ft	15/19

NR = Not Required.

For SI: 1 foot = 304.8 mm.

a. *R*-values are minimums. *U*-factors and SHGC are maximums. Where insulation is installed in a cavity that is less than the label or design thickness of the insulation, the installed *R*-value of the insulation shall be not less than the *R*-value specified in the table.

b. The fenestration *U*-factor column excludes skylights. The SHGC column applies to all glazed fenestration.
Exception: In Climate Zones 1 through 3, skylights shall be permitted to be excluded from glazed fenestration SHGC requirements provided that the SHGC for such skylights does not exceed 0.30.

c. "10/13" means R-10 continuous insulation on the interior or exterior of the home or R-13 cavity insulation on the interior of the basement wall.
"15/19" means R-15 continuous insulation on the interior or exterior of the home or R-19 cavity insulation at the interior of the basement wall. Alternatively, compliance with "15/19" shall be R-13 cavity insulation on the interior of the basement wall plus R-5 continuous insulation on the interior or exterior of the home.

d. R-5 insulation shall be provided under the full slab area of a heated slab in addition to the required slab edge insulation *R*-value for slabs. as indicated in the table. The slab edge insulation for heated slabs shall not be required to extend below the slab.

e. There are no SHGC requirements in the Marine Zone.

f. Basement wall insulation is not required in warm-humid locations as defined by Figure R301.1 and Table R301.1.

g. Alternatively, insulation sufficient to fill the framing cavity and providing not less than an *R*-value of R-19.

h. The first value is cavity insulation, the second value is continuous insulation. Therefore, as an example, "13+5" means R-13 cavity insulation plus R-5 continuous insulation.

i. Mass walls shall be in accordance with Section R402.2.5. The second *R*-value applies where more than half of the insulation is on the interior of the mass wall.

TABLE R402.1.4
EQUIVALENT *U*-FACTORS[a]

CLIMATE ZONE	FENESTRATION *U*-FACTOR	SKYLIGHT *U*-FACTOR	CEILING *U*-FACTOR	FRAME WALL *U*-FACTOR	MASS WALL *U*-FACTOR[b]	FLOOR *U*-FACTOR	BASEMENT WALL *U*-FACTOR	CRAWL SPACE WALL *U*-FACTOR
1	0.50	0.75	0.035	0.084	0.197	0.064	0.360	0.477
2	0.40	0.65	0.030	0.084	0.165	0.064	0.360	0.477
3	0.32	0.55	0.030	0.060	0.098	0.047	0.091[c]	0.136
4 except Marine	0.32	0.55	0.026	0.060	0.098	0.047	0.059	0.065
5 and Marine 4	0.30	0.55	0.026	0.060	0.082	0.033	0.050	0.055
6	0.30	0.55	0.026	0.045	0.060	0.033	0.050	0.055
7 and 8	0.30	0.55	0.026	0.045	0.057	0.028	0.050	0.055

a. Nonfenestration *U*-factors shall be obtained from measurement, calculation or an approved source.

b. Mass walls shall be in accordance with Section R402.2.5. Where more than half the insulation is on the interior, the mass wall *U*-factors shall not exceed 0.17 in Climate Zone 1, 0.14 in Climate Zone 2, 0.12 in Climate Zone 3, 0.087 in Climate Zone 4 except Marine, 0.065 in Climate Zone 5 and Marine 4, and 0.057 in Climate Zones 6 through 8.

c. In warm-humid locations as defined by Figure R301.1 and Table R301.1, the basement wall *U*-factor shall not exceed 0.360.

R402.1.5 Total UA alternative. Where the total *building thermal envelope* UA, the sum of *U*-factor times assembly area, is less than or equal to the total UA resulting from multiplying the *U*-factors in Table R402.1.4 by the same assembly area as in the proposed *building*, the *building* shall be considered to be in compliance with Table R402.1.2. The UA calculation shall be performed using a method consistent with the ASHRAE *Handbook of Fundamentals* and shall include the thermal bridging effects of framing materials. In addition to UA compliance, the SHGC requirements shall be met.

R402.2 Specific insulation requirements (Prescriptive). In addition to the requirements of Section R402.1, insulation shall meet the specific requirements of Sections R402.2.1 through R402.2.13.

R402.2.1 Ceilings with attic spaces. Where Section R402.1.2 requires R-38 insulation in the ceiling, installing R-30 over 100 percent of the ceiling area requiring insulation shall satisfy the requirement for R-38 wherever the full height of uncompressed R-30 insulation extends over the wall top plate at the eaves. Where Section R402.1.2 requires R-49 insulation in the ceiling, installing R-38 over 100 percent of the ceiling area requiring insulation shall satisfy the requirement for R-49 insulation wherever the full height of uncompressed R-38 insulation extends over the wall top plate at the eaves. This reduction shall not apply to the *U*-factor alternative approach in Section R402.1.4 and the Total UA alternative in Section R402.1.5.

R402.2.2 Ceilings without attic spaces. Where Section R402.1.2 requires insulation *R*-values greater than R-30 in the ceiling and the design of the roof/ceiling assembly does not allow sufficient space for the required insulation, the minimum required insulation *R*-value for such roof/ceiling assemblies shall be R-30. Insulation shall extend over the top of the wall plate to the outer edge of such plate and shall not be compressed. This reduction of insulation from the requirements of Section R402.1.2 shall be limited to 500 square feet (46 m^2) or 20 percent of the total insulated ceiling area, whichever is less. This reduction shall not apply to the *U*-factor alternative approach in Section R402.1.4 and the Total UA alternative in Section R402.1.5.

R402.2.3 Eave baffle. For air-permeable insulations in vented attics, a baffle shall be installed adjacent to soffit and eave vents. Baffles shall maintain an opening equal or greater than the size of the vent. The baffle shall extend over the top of the attic insulation. The baffle shall be permitted to be any solid material.

R402.2.4 Access hatches and doors. Access doors from *conditioned spaces* to *unconditioned spaces* such as attics and crawl spaces shall be weatherstripped and insulated to a level equivalent to the insulation on the surrounding surfaces. Access that prevents damaging or compressing the insulation shall be provided to all equipment. Where loose-fill insulation is installed, a wood-framed or equivalent baffle or retainer shall be installed to prevent the loose-fill insulation from spilling into the living space when the attic access is opened. The baffle or retainer shall provide a permanent means of maintaining the installed *R*-value of the loose-fill insulation.

Exception: Vertical doors providing access from *conditioned spaces* to *unconditioned spaces* that comply with the fenestration requirements of Table R402.1.2 based on the applicable *climate zone* specified in Chapter 3.

R402.2.5 Mass walls. Mass walls where used as a component of the *building thermal envelope* shall be one of the following:

1. Above-ground walls of concrete block, concrete, insulated concrete form, masonry cavity, brick but not brick veneer, adobe, compressed earth block, rammed earth, solid timber or solid logs.
2. Any wall having a heat capacity greater than or equal to 6 Btu/ft² • °F (123 kJ/m² • K).

R402.2.6 Steel-frame ceilings, walls and floors. Steel-frame ceilings, walls, and floors shall comply with the

**TABLE R402.2.6
STEEL-FRAME CEILING, WALL
AND FLOOR INSULATION *R*-VALUES**

WOOD FRAME *R*-VALUE REQUIREMENT	COLD-FORMED STEEL-FRAME EQUIVALENT *R*-VALUE[a]
Steel Truss Ceilings[b]	
R-30	R-38 or R-30 + 3 or R-26 + 5
R-38	R-49 or R-38 + 3
R-49	R-38 + 5
Steel Joist Ceilings[b]	
R-30	R-38 in 2 × 4 or 2 × 6 or 2 × 8 R-49 in any framing
R-38	R-49 in 2 × 4 or 2 × 6 or 2 × 8 or 2 × 10
Steel-Framed Wall, 16 inches on center	
R-13	R-13 + 4.2 or R-21 + 2.8 or R-0 + 9.3 or R-15 + 3.8 or R-21 + 3.1
R-13 + 3	R-0 + 11.2 or R-13 + 6.1 or R-15 + 5.7 or R-19 + 5.0 or R-21 + 4.7
R-20	R-0 + 14.0 or R-13 + 8.9 or R-15 + 8.5 or R-19 + 7.8 or R-19 + 6.2 or R-21 + 7.5
R-20 + 5	R-13 + 12.7 or R-15 + 12.3 or R-19 + 11.6 or R-21 + 11.3 or R-25 + 10.9
R-21	R-0 + 14.6 or R-13 + 9.5 or R-15 + 9.1 or R-19 + 8.4 or R-21 + 8.1 or R-25 + 7.7
Steel Framed Wall, 24 inches on center	
R-13	R-0 + 9.3 or R-13 + 3.0 or R-15 + 2.4
R-13 + 3	R-0 + 11.2 or R-13 + 4.9 or R-15 + 4.3 or R-19 + 3.5 or R-21 + 3.1
R-20	R-0 + 14.0 or R-13 + 7.7 or R-15 + 7.1 or R-19 + 6.3 or R-21 + 5.9
R-20 + 5	R-13 + 11.5 or R-15 + 10.9 or R-19 + 10.1 or R-21 + 9.7 or R-25 + 9.1
R-21	R-0 + 14.6 or R-13 + 8.3 or R-15 + 7.7 or R-19 + 6.9 or R-21 + 6.5 or R-25 + 5.9
Steel Joist Floor	
R-13	R-19 in 2 × 6, or R-19 + 6 in 2 × 8 or 2 × 10
R-19	R-19 + 6 in 2 × 6, or R-19 + 12 in 2 × 8 or 2 × 10

a. The first value is cavity insulation *R*-value, the second value is continuous insulation *R*-value. Therefore, for example, "R-30+3" means R-30 cavity insulation plus R-3 continuous insulation.

b. Insulation exceeding the height of the framing shall cover the framing.

insulation requirements of Table R402.2.6 or the *U*-factor requirements of Table R402.1.4. The calculation of the *U*-factor for a steel-frame envelope assembly shall use a series-parallel path calculation method.

R402.2.7 Walls with partial structural sheathing. Where Section R402.1.2 requires continuous insulation on *exterior walls* and structural sheathing covers 40 percent or less of the gross area of all *exterior walls*, the required continuous insulation *R*-value shall be permitted to be reduced by an amount necessary, but not more than R-3 to result in a consistent total sheathing thickness on areas of the walls covered by structural sheathing. This reduction shall not apply to the *U*-factor alternative in Section R402.1.4 and the Total UA alternative in Section R402.1.5.

R402.2.8 Floors. Floor framing-*cavity insulation* shall be installed to maintain permanent contact with the underside of the subfloor decking.

> **Exception:** As an alternative, the floor framing-*cavity insulation* shall be in contact with the topside of sheathing or continuous insulation installed on the bottom side of floor framing where combined with insulation that meets or exceeds the minimum wood frame wall *R*-value in Table R402.1.2 and that extends from the bottom to the top of all perimeter floor framing members.

R402.2.9 Basement walls. Walls associated with conditioned basements shall be insulated from the top of the *basement wall* down to 10 feet (3048 mm) below grade or to the basement floor, whichever is less. Walls associated with unconditioned basements shall comply with this requirement except where the floor overhead is insulated in accordance with Sections R402.1.2 and R402.2.8.

R402.2.10 Slab-on-grade floors. Slab-on-grade floors with a floor surface less than 12 inches (305 mm) below grade shall be insulated in accordance with Table R402.1.2. The insulation shall extend downward from the top of the slab on the outside or inside of the foundation wall. Insulation located below grade shall be extended the distance provided in Table R402.1.2 by any combination of vertical insulation, insulation extending under the slab or insulation extending out from the *building*. Insulation extending away from the *building* shall be protected by pavement or by not less than 10 inches (254 mm) of soil. The top edge of the insulation installed between the *exterior wall* and the edge of the interior slab shall be permitted to be cut at a 45-degree (0.79 rad) angle away from the *exterior wall*. Slab-edge insulation is not required in jurisdictions designated by the *code official* as having a very heavy termite infestation.

R402.2.11 Crawl space walls. As an alternative to insulating floors over crawl spaces, crawl space walls shall be insulated provided that the crawl space is not vented to the outdoors. Crawl space wall insulation shall be permanently fastened to the wall and shall extend downward from the floor to the finished grade elevation and then vertically or horizontally for not less than an additional 24 inches (610 mm). Exposed earth in unvented crawl space foundations shall be covered with a continuous Class I vapor retarder in accordance with the *International Building Code* or *International Residential Code*, as applicable. Joints of the vapor retarder shall overlap by 6 inches (153 mm) and be sealed or taped. The edges of the vapor retarder shall extend not less than 6 inches (153 mm) up stem walls and shall be attached to the stem walls.

R402.2.12 Masonry veneer. Insulation shall not be required on the horizontal portion of a foundation that supports a masonry veneer.

R402.2.13 Sunroom insulation. *Sunrooms* enclosing *conditioned space* shall meet the insulation requirements of this code.

> **Exception:** For *sunrooms* with *thermal isolation*, and enclosing *conditioned space*, the following exceptions to the insulation requirements of this code shall apply:
>
> 1. The minimum ceiling insulation *R*-values shall be R-19 in *Climate Zones* 1 through 4 and R-24 in *Climate Zones* 5 through 8.
> 2. The minimum wall insulation *R*-value shall be R-13 in all *climate zones*. Walls separating a *sunroom* with a *thermal isolation* from *conditioned space* shall comply with the *building thermal envelope* requirements of this code.

R402.3 Fenestration (Prescriptive). In addition to the requirements of Section R402, fenestration shall comply with Sections R402.3.1 through R402.3.5.

R402.3.1 *U*-factor. An area-weighted average of fenestration products shall be permitted to satisfy the *U*-factor requirements.

R402.3.2 Glazed fenestration SHGC. An area-weighted average of fenestration products more than 50-percent glazed shall be permitted to satisfy the SHGC requirements.

Dynamic glazing shall be permitted to satisfy the SHGC requirements of Table R402.1.2 provided that the ratio of the higher to lower labeled SHGC is greater than or equal to 2.4, and the dynamic glazing is automatically controlled to modulate the amount of solar gain into the space in multiple steps. Dynamic glazing shall be considered separately from other fenestration, and area-weighted averaging with other fenestration that is not dynamic glazing shall be prohibited.

> **Exception:** Dynamic glazing shall not be required to comply with this section where both the lower and higher labeled SHGC comply with the requirements of Table R402.1.2.

R402.3.3 Glazed fenestration exemption. Not greater than 15 square feet (1.4 m^2) of glazed fenestration per dwelling unit shall be exempt from the *U*-factor and SHGC requirements in Section R402.1.2. This exemption shall not apply to the *U*-factor alternative in Section R402.1.4 and the Total UA alternative in Section R402.1.5.

R402.3.4 Opaque door exemption. One side-hinged opaque door assembly not greater than 24 square feet (2.22 m^2) in area shall be exempt from the *U*-factor requirement in Section R402.1.2. This exemption shall not apply to the *U*-factor alternative in Section R402.1.4 and the Total UA alternative in Section R402.1.5.

R402.3.5 Sunroom fenestration. *Sunrooms* enclosing *conditioned space* shall comply with the fenestration requirements of this code.

Exception: In *Climate Zones* 2 through 8, for *sunrooms* with *thermal isolation* and enclosing *conditioned space*, the fenestration *U*-factor shall not exceed 0.45 and the *skylight U*-factor shall not exceed 0.70.

New fenestration separating the *sunroom* with *thermal isolation* from *conditioned space* shall comply with the *building thermal envelope* requirements of this code.

R402.4 Air leakage (Mandatory). The *building thermal envelope* shall be constructed to limit air leakage in accordance with the requirements of Sections R402.4.1 through R402.4.5.

R402.4.1 Building thermal envelope. The *building thermal envelope* shall comply with Sections R402.4.1.1 and R402.4.1.2. The sealing methods between dissimilar materials shall allow for differential expansion and contraction.

R402.4.1.1 Installation. The components of the *building thermal envelope* as indicated in Table R402.4.1.1 shall be installed in accordance with the manufacturer's instructions and the criteria indicated in Table R402.4.1.1, as applicable to the method of construction. Where required by the *code official*, an *approved* third party shall inspect all components and verify compliance.

R402.4.1.2 Testing. The *building* or dwelling unit shall be tested and verified as having an air leakage rate not exceeding five air changes per hour in *Climate Zones* 1 and 2, and three air changes per hour in *Climate Zones* 3 through 8. Testing shall be conducted in accordance with RESNET/ICC 380, ASTM E779 or ASTM E1827 and reported at a pressure of 0.2 inch w.g. (50 Pascals). Where required by the *code official*, testing shall be conducted by an *approved* third party. A written report of the results of the test shall be signed by the party conducting the test and provided to the *code official*. Testing shall be performed at any time after creation of all penetrations of the *building thermal envelope*.

During testing:

1. Exterior windows and doors, fireplace and stove doors shall be closed, but not sealed, beyond the intended weatherstripping or other infiltration control measures.
2. Dampers including exhaust, intake, makeup air, backdraft and flue dampers shall be closed, but not sealed beyond intended infiltration control measures.
3. Interior doors, where installed at the time of the test, shall be open.
4. Exterior or interior terminations for continuous ventilation systems shall be sealed.
5. Heating and cooling systems, where installed at the time of the test, shall be turned off.
6. Supply and return registers, where installed at the time of the test, shall be fully open.

R402.4.2 Fireplaces. New wood-burning fireplaces shall have tight-fitting flue dampers or doors, and outdoor combustion air. Where using tight-fitting doors on factory-built fireplaces *listed* and *labeled* in accordance with UL 127, the doors shall be tested and *listed* for the fireplace.

R402.4.3 Fenestration air leakage. Windows, *skylights* and sliding glass doors shall have an air infiltration rate of not greater than 0.3 cfm per square foot (1.5 L/s/m^2), and for swinging doors, not greater than 0.5 cfm per square foot (2.6 L/s/m^2), when tested in accordance with NFRC 400 or AAMA/WDMA/CSA 101/I.S.2/A440 by an accredited, independent laboratory and *listed* and labeled by the manufacturer.

Exception: Site-built windows, *skylights* and doors.

R402.4.4 Rooms containing fuel-burning appliances. In *Climate Zones* 3 through 8, where open combustion air ducts provide combustion air to open combustion fuel burning appliances, the appliances and combustion air opening shall be located outside the *building thermal envelope* or enclosed in a room that is isolated from inside the thermal envelope. Such rooms shall be sealed and insulated in accordance with the envelope requirements of Table R402.1.2, where the walls, floors and ceilings shall meet not less than the *basement wall R*-value requirement. The door into the room shall be fully gasketed and any water lines and ducts in the room insulated in accordance with Section R403. The combustion air duct shall be insulated where it passes through *conditioned space* to an *R*-value of not less than R-8.

Exceptions:

1. Direct vent appliances with both intake and exhaust pipes installed continuous to the outside.
2. Fireplaces and stoves complying with Section R402.4.2 and Section R1006 of the *International Residential Code*.

R402.4.5 Recessed lighting. Recessed luminaires installed in the *building thermal envelope* shall be sealed to limit air leakage between conditioned and *unconditioned spaces*. Recessed luminaires shall be IC-rated and *labeled* as having an air leakage rate of not greater than 2.0 cfm (0.944 L/s) when tested in accordance with ASTM E283 at a pressure differential of 1.57 psf (75 Pa). Recessed luminaires shall be sealed with a gasket or caulked between the housing and the interior wall or ceiling covering.

R402.5 Maximum fenestration *U*-factor and SHGC (Mandatory). The area-weighted average maximum fenestration *U*-factor permitted using tradeoffs from Section R402.1.5 or R405 shall be 0.48 in *Climate Zones* 4 and 5 and 0.40 in *Climate Zones* 6 through 8 for vertical fenestration, and 0.75 in *Climate Zones* 4 through 8 for *skylights*. The area-weighted average maximum fenestration SHGC permitted using tradeoffs from Section R405 in *Climate Zones* 1 through 3 shall be 0.50.

**TABLE R402.4.1.1
AIR BARRIER AND INSULATION INSTALLATION[a]**

COMPONENT	AIR BARRIER CRITERIA	INSULATION INSTALLATION CRITERIA
General requirements	A continuous air barrier shall be installed in the building envelope. The exterior thermal envelope contains a continuous air barrier. Breaks or joints in the air barrier shall be sealed.	Air-permeable insulation shall not be used as a sealing material.
Ceiling/attic	The air barrier in any dropped ceiling or soffit shall be aligned with the insulation and any gaps in the air barrier shall be sealed. Access openings, drop down stairs or knee wall doors to unconditioned attic spaces shall be sealed.	The insulation in any dropped ceiling/soffit shall be aligned with the air barrier.
Walls	The junction of the foundation and sill plate shall be sealed. The junction of the top plate and the top of exterior walls shall be sealed. Knee walls shall be sealed.	Cavities within corners and headers of frame walls shall be insulated by completely filling the cavity with a material having a thermal resistance, *R*-value, of not less than R-3 per inch. Exterior thermal envelope insulation for framed walls shall be installed in substantial contact and continuous alignment with the air barrier.
Windows, skylights and doors	The space between framing and skylights, and the jambs of windows and doors, shall be sealed.	—
Rim joists	Rim joists shall include the air barrier.	Rim joists shall be insulated.
Floors, including cantilevered floors and floors above garages	The air barrier shall be installed at any exposed edge of insulation.	Floor framing cavity insulation shall be installed to maintain permanent contact with the underside of subfloor decking. Alternatively, floor framing cavity insulation shall be in contact with the top side of sheathing, or continuous insulation installed on the underside of floor framing; and shall extend from the bottom to the top of all perimeter floor framing members.
Crawl space walls	Exposed earth in unvented crawl spaces shall be covered with a Class I vapor retarder with overlapping joints taped.	Crawl space insulation, where provided instead of floor insulation, shall be permanently attached to the walls.
Shafts, penetrations	Duct shafts, utility penetrations, and flue shafts opening to exterior or unconditioned space shall be sealed.	—
Narrow cavities	—	Batts to be installed in narrow cavities shall be cut to fit or narrow cavities shall be filled with insulation that on installation readily conforms to the available cavity space.
Garage separation	Air sealing shall be provided between the garage and conditioned spaces.	—
Recessed lighting	Recessed light fixtures installed in the building thermal envelope shall be sealed to the finished surface.	Recessed light fixtures installed in the building thermal envelope shall be air tight and IC rated.
Plumbing and wiring	—	In exterior walls, batt insulation shall be cut neatly to fit around wiring and plumbing, or insulation, that on installation readily conforms to available space, shall extend behind piping and wiring.
Shower/tub on exterior wall	The air barrier installed at exterior walls adjacent to showers and tubs shall separate the wall from the shower or tub.	Exterior walls adjacent to showers and tubs shall be insulated.
Electrical/phone box on exterior walls	The air barrier shall be installed behind electrical and communication boxes. Alternatively, air-sealed boxes shall be installed.	—
HVAC register boots	HVAC supply and return register boots that penetrate building thermal envelope shall be sealed to the subfloor, wall covering or ceiling penetrated by the boot.	—
Concealed sprinklers	Where required to be sealed, concealed fire sprinklers shall only be sealed in a manner that is recommended by the manufacturer. Caulking or other adhesive sealants shall not be used to fill voids between fire sprinkler cover plates and walls or ceilings.	—

a. Inspection of log walls shall be in accordance with the provisions of ICC 400.

SECTION R403
SYSTEMS

R403.1 Controls (Mandatory). Not less than one thermostat shall be provided for each separate heating and cooling system.

R403.1.1 Programmable thermostat. The thermostat controlling the primary heating or cooling system of the dwelling unit shall be capable of controlling the heating and cooling system on a daily schedule to maintain different temperature setpoints at different times of the day. This thermostat shall include the capability to set back or temporarily operate the system to maintain *zone* temperatures of not less than 55°F (13°C) to not greater than 85°F (29°C). The thermostat shall be programmed initially by the manufacturer with a heating temperature setpoint of not greater than 70°F (21°C) and a cooling temperature setpoint of not less than 78°F (26°C).

R403.1.2 Heat pump supplementary heat (Mandatory). Heat pumps having supplementary electric-resistance heat shall have controls that, except during defrost, prevent supplemental heat operation when the heat pump compressor can meet the heating load.

R403.2 Hot water boiler outdoor temperature setback. Hot water boilers that supply heat to the *building* through one- or two-pipe heating systems shall have an outdoor setback control that decreases the boiler water temperature based on the outdoor temperature.

R403.3 Ducts. Ducts and air handlers shall be installed in accordance with Sections R403.3.1 through R403.3.7.

R403.3.1 Insulation (Prescriptive). Supply and return ducts in attics shall be insulated to an *R*-value of not less than R-8 for ducts 3 inches (76 mm) in diameter and larger and not less than R-6 for ducts smaller than 3 inches (76 mm) in diameter. Supply and return ducts in other portions of the *building* shall be insulated to not less than R-6 for ducts 3 inches (76 mm) in diameter and not less than R-4.2 for ducts smaller than 3 inches (76 mm) in diameter.

Exception: Ducts or portions thereof located completely inside the *building thermal envelope.*

R403.3.2 Sealing (Mandatory). Ducts, air handlers and filter boxes shall be sealed. Joints and seams shall comply with either the *International Mechanical Code* or *International Residential Code*, as applicable.

➡ **R403.3.2.1 Sealed air handler.** Air handlers shall have a manufacturer's designation for an air leakage of not greater than 2 percent of the design airflow rate when tested in accordance with ASHRAE 193.

R403.3.3 Duct testing (Mandatory). Ducts shall be pressure tested to determine air leakage by one of the following methods:

1. Rough-in test: Total leakage shall be measured with a pressure differential of 0.1 inch w.g. (25 Pa) across the system, including the manufacturer's air handler enclosure if installed at the time of the test. Registers shall be taped or otherwise sealed during the test.
2. Postconstruction test: Total leakage shall be measured with a pressure differential of 0.1 inch w.g. (25 Pa) across the entire system, including the manufacturer's air handler enclosure. Registers shall be taped or otherwise sealed during the test.

Exceptions:

1. A duct air-leakage test shall not be required where the ducts and air handlers are located entirely within the *building thermal envelope*.
2. A duct air-leakage test shall not be required for ducts serving heat or energy recovery ventilators that are not integrated with ducts serving heating or cooling systems.

A written report of the results of the test shall be signed by the party conducting the test and provided to the *code official*.

R403.3.4 Duct leakage (Prescriptive). The total leakage of the ducts, where measured in accordance with Section R403.3.3, shall be as follows:

1. Rough-in test: The total leakage shall be less than or equal to 4 cubic feet per minute (113.3 L/min) per 100 square feet (9.29 m^2) of conditioned floor area where the air handler is installed at the time of the test. Where the air handler is not installed at the time of the test, the total leakage shall be less than or equal to 3 cubic feet per minute (85 L/min) per 100 square feet (9.29 m^2) of conditioned floor area.
2. Postconstruction test: Total leakage shall be less than or equal to 4 cubic feet per minute (113.3 L/min) per 100 square feet (9.29 m^2) of conditioned floor area.

R403.3.5 Building cavities (Mandatory). *Building* framing cavities shall not be used as ducts or plenums.

R403.3.6 Ducts buried within ceiling insulation. Where supply and return air ducts are partially or completely buried in ceiling insulation, such ducts shall comply with all of the following:

1. The supply and return ducts shall have an insulation *R*-value not less than R-8.
2. At all points along each duct, the sum of the ceiling insulation *R*-value against and above the top of the duct, and against and below the bottom of the duct, shall be not less than R-19, excluding the *R*-value of the duct insulation.
3. In *Climate Zones* 1A, 2A and 3A, the supply ducts shall be completely buried within ceiling insulation, insulated to an *R*-value of not less than R-13 and in compliance with the vapor retarder requirements of Section 604.11 of the *International Mechanical Code* or Section M1601.4.6 of the *International Residential Code*, as applicable.

Exception: Sections of the supply duct that are less than 3 feet (914 mm) from the supply outlet shall not be required to comply with these requirements.

R403.3.6.1 Effective *R*-value of deeply buried ducts. Where using a simulated energy performance analysis, sections of ducts that are: installed in accordance with Section R403.3.6; located directly on, or within 5.5 inches (140 mm) of the ceiling; surrounded with blown-in attic insulation having an *R*-value of R-30 or greater and located such that the top of the duct is not less than 3.5 inches (89 mm) below the top of the insulation, shall be considered as having an effective duct insulation *R*-value of R-25.

R403.3.7 Ducts located in conditioned space. For ducts to be considered as inside a conditioned space, such ducts shall comply with either of the following:

1. The duct system shall be located completely within the continuous air barrier and within the building thermal envelope.
2. The ducts shall be buried within ceiling insulation in accordance with Section R403.3.6 and all of the following conditions shall exist:
   2.1. The air handler is located completely within the *continuous air barrier* and within the building thermal envelope.
   2.2. The duct leakage, as measured either by a rough-in test of the ducts or a post-construction total system leakage test to outside the building thermal envelope in accordance with Section R403.3.4, is less than or equal to 1.5 cubic feet per minute (42.5 L/min) per 100 square feet (9.29 m^2) of conditioned floor area served by the duct system.
   2.3. The ceiling insulation *R*-value installed against and above the insulated duct is greater than or equal to the proposed ceiling insulation *R*-value, less the *R*-value of the insulation on the duct.

R403.4 Mechanical system piping insulation (Mandatory). Mechanical system piping capable of carrying fluids greater than 105°F (41°C) or less than 55°F (13°C) shall be insulated to an *R*-value of not less than R-3.

R403.4.1 Protection of piping insulation. Piping insulation exposed to weather shall be protected from damage, including that caused by sunlight, moisture, equipment maintenance and wind. The protection shall provide shielding from solar radiation that can cause degradation of the material. Adhesive tape shall be prohibited.

R403.5 Service hot water systems. Energy conservation measures for service hot water systems shall be in accordance with Sections R403.5.1 through R403.5.4.

R403.5.1 Heated water circulation and temperature maintenance systems (Mandatory). Heated water circulation systems shall be in accordance with Section R403.5.1.1. Heat trace temperature maintenance systems shall be in accordance with Section R403.5.1.2. Automatic controls, temperature sensors and pumps shall be *accessible*. Manual controls shall be readily *accessible*.

R403.5.1.1 Circulation systems. Heated water circulation systems shall be provided with a circulation pump. The system return pipe shall be a dedicated return pipe or a cold water supply pipe. Gravity and thermosyphon circulation systems shall be prohibited. Controls for circulating hot water system pumps shall start the pump based on the identification of a demand for hot water within the occupancy. The controls shall automatically turn off the pump when the water in the circulation loop is at the desired temperature and when there is no demand for hot water.

R403.5.1.2 Heat trace systems. Electric heat trace systems shall comply with IEEE 515.1 or UL 515. Controls for such systems shall automatically adjust the energy input to the heat tracing to maintain the desired water temperature in the piping in accordance with the times when heated water is used in the occupancy.

R403.5.2 Demand recirculation water systems. *Demand recirculation water systems* shall have controls that comply with both of the following:

1. The controls shall start the pump upon receiving a signal from the action of a user of a fixture or appliance, sensing the presence of a user of a fixture or sensing the flow of hot or tempered water to a fixture fitting or appliance.
2. The controls shall limit the temperature of the water entering the cold water piping to not greater than 104°F (40°C).

R403.5.3 Hot water pipe insulation (Prescriptive). Insulation for hot water piping with a thermal resistance, *R*-value, of not less than R-3 shall be applied to the following:

1. Piping $^3/_4$ inch (19.1 mm) and larger in nominal diameter.
2. Piping serving more than one dwelling unit.
3. Piping located outside the *conditioned space.*
4. Piping from the water heater to a distribution manifold.
5. Piping located under a floor slab.
6. Buried piping.
7. Supply and return piping in recirculation systems other than demand recirculation systems.

R403.5.4 Drain water heat recovery units. Drain water heat recovery units shall comply with CSA B55.2. Drain water heat recovery units shall be tested in accordance with CSA B55.1. Potable water-side pressure loss of drain water heat recovery units shall be less than 3 psi (20.7 kPa) for individual units connected to one or two showers. Potable water-side pressure loss of drain water heat recovery units shall be less than 2 psi (13.8 kPa) for individual units connected to three or more showers.

R403.6 Mechanical ventilation (Mandatory). The *building* shall be provided with ventilation that complies with the requirements of the *International Residential Code* or *International Mechanical Code*, as applicable, or with other *approved* means of ventilation. Outdoor air intakes and exhausts shall have automatic or gravity dampers that close when the ventilation system is not operating.

R403.6.1 Whole-house mechanical ventilation system fan efficacy. Fans used to provide whole-house mechanical ventilation shall meet the efficacy requirements of Table R403.6.1.

Exception: Where an air handler that is integral to tested and *listed* HVAC equipment is used to provide whole-house mechanical ventilation, the air handler shall be powered by an electronically commutated motor.

R403.7 Equipment sizing and efficiency rating (Mandatory). Heating and cooling equipment shall be sized in accordance with ACCA Manual S based on *building* loads calculated in accordance with ACCA Manual J or other *approved* heating and cooling calculation methodologies. New or replacement heating and cooling equipment shall have an efficiency rating equal to or greater than the minimum required by federal law for the geographic location where the equipment is installed.

R403.8 Systems serving multiple dwelling units (Mandatory). Systems serving multiple dwelling units shall comply with Sections C403 and C404 of the *International Energy Conservation Code*—Commercial Provisions instead of Section R403.

R403.9 Snow melt and ice system controls (Mandatory). Snow- and ice-melting systems, supplied through energy service to the building, shall include automatic controls capable of shutting off the system when the pavement temperature is greater than 50°F (10°C) and precipitation is not falling, and an automatic or manual control that will allow shutoff when the outdoor temperature is greater than 40°F (4.8°C).

R403.10 Pools and permanent spa energy consumption (Mandatory). The energy consumption of pools and permanent spas shall be in accordance with Sections R403.10.1 through R403.10.3.

R403.10.1 Heaters. The electric power to heaters shall be controlled by a readily *accessible* on-off switch that is an integral part of the heater mounted on the exterior of the heater, or external to and within 3 feet (914 mm) of the heater. Operation of such switch shall not change the setting of the heater thermostat. Such switches shall be in addition to a circuit breaker for the power to the heater. Gas-fired heaters shall not be equipped with continuously burning ignition pilots.

R403.10.2 Time switches. Time switches or other control methods that can automatically turn off and on according to a preset schedule shall be installed for heaters and pump motors. Heaters and pump motors that have built-in time switches shall be in compliance with this section.

Exceptions:

1. Where public health standards require 24-hour pump operation.
2. Pumps that operate solar- and waste-heat-recovery pool heating systems.

R403.10.3 Covers. Outdoor heated pools and outdoor permanent spas shall be provided with a vapor-retardant cover or other *approved* vapor-retardant means.

Exception: Where more than 75 percent of the energy for heating, computed over an operation season of not less than three calendar months, is from a heat pump or an on-site renewable energy system, covers or other vapor-retardant means shall not be required.

R403.11 Portable spas (Mandatory). The energy consumption of electric-powered portable spas shall be controlled by the requirements of APSP 14.

R403.12 Residential pools and permanent residential spas. Residential swimming pools and permanent residential spas that are accessory to detached one- and two-family dwellings and townhouses three stories or less in height above grade plane and that are available only to the household and its guests shall be in accordance with APSP 15.

SECTION R404 ELECTRICAL POWER AND LIGHTING SYSTEMS

R404.1 Lighting equipment (Mandatory). Not less than 90 percent of the permanently installed lighting fixtures shall contain only high-efficacy lamps.

R404.1.1 Lighting equipment (Mandatory). Fuel gas lighting systems shall not have continuously burning pilot lights.

SECTION R405 SIMULATED PERFORMANCE ALTERNATIVE (PERFORMANCE)

R405.1 Scope. This section establishes criteria for compliance using simulated energy performance analysis. Such

TABLE R403.6.1
WHOLE-HOUSE MECHANICAL VENTILATION SYSTEM FAN EFFICACY[a]

FAN LOCATION	AIR FLOW RATE MINIMUM (CFM)	MINIMUM EFFICACY (CFM/WATT)	AIR FLOW RATE MAXIMUM (CFM)
HRV or ERV	Any	1.2 cfm/watt	Any
Range hoods	Any	2.8 cfm/watt	Any
In-line fan	Any	2.8 cfm/watt	Any
Bathroom, utility room	10	1.4 cfm/watt	< 90
Bathroom, utility room	90	2.8 cfm/watt	Any

For SI: 1 cfm = 28.3 L/min.

a. When tested in accordance with HVI Standard 916.

analysis shall include heating, cooling, mechanical ventilation and service water heating energy only.

R405.2 Mandatory requirements. Compliance with this section requires that the mandatory provisions identified in Section R401.2 be met. Supply and return ducts not completely inside the *building thermal envelope* shall be insulated to an *R*-value of not less than R-6.

R405.3 Performance-based compliance. Compliance based on simulated energy performance requires that a proposed residence (*proposed design)* be shown to have an annual energy cost that is less than or equal to the annual energy cost of the *standard reference design.* Energy prices shall be taken from a source *approved* by the *code official,* such as the Department of Energy, Energy Information Administration's State Energy Data System Prices and Expenditures reports. *Code officials* shall be permitted to require time-of-use pricing in energy cost calculations.

Exception: The energy use based on source energy expressed in Btu or Btu per square foot of *conditioned floor area* shall be permitted to be substituted for the energy cost. The source energy multiplier for electricity shall be 3.16. The source energy multiplier for fuels other than electricity shall be 1.1.

R405.4 Documentation. Documentation of the software used for the performance design and the parameters for the *building* shall be in accordance with Sections R405.4.1 through R405.4.3.

R405.4.1 Compliance software tools. Documentation verifying that the methods and accuracy of the compliance software tools conform to the provisions of this section shall be provided to the *code official.*

R405.4.2 Compliance report. Compliance software tools shall generate a report that documents that the *proposed design* complies with Section R405.3. A compliance report on the *proposed design* shall be submitted with the application for the *building* permit. Upon completion of the *building*, a compliance report based on the as-built condition of the *building* shall be submitted to the *code official* before a certificate of occupancy is issued. Batch sampling of *buildings* to determine energy code compliance shall only be allowed for stacked multiple-family units.

Compliance reports shall include information in accordance with Sections R405.4.2.1 and R405.4.2.2. Where the *proposed design* of a *building* could be built on different sites where the cardinal orientation of the building on each site is different, compliance of the *proposed design* for the purposes of the application for the building permit shall be based on the worst-case orientation, worst-case configuration, worst-case *building* air leakage and worst- case duct leakage. Such worst-case parameters shall be used as inputs to the compliance software for energy analysis.

R405.4.2.1 Compliance report for permit application. A compliance report submitted with the application for building permit shall include the following:

1. Building street address, or other building site identification.
2. A statement indicating that the *proposed design* complies with Section R405.3.
3. An inspection checklist documenting the building component characteristics of the *proposed design* as indicated in Table R405.5.2(1). The inspection checklist shall show results for both the *standard reference design* and the *proposed design* with user inputs to the compliance software to generate the results.
4. A site-specific energy analysis report that is in compliance with Section R405.3.
5. The name of the individual performing the analysis and generating the report.
6. The name and version of the compliance software tool.

R405.4.2.2 Compliance report for certificate of occupancy. A compliance report submitted for obtaining the certificate of occupancy shall include the following:

1. Building street address, or other building site identification.
2. A statement indicating that the as-built building complies with Section R405.3.
3. A certificate indicating that the building passes the performance matrix for code compliance and indicating the energy saving features of the buildings.
4. A site-specific energy analysis report that is in compliance with Section R405.3.
5. The name of the individual performing the analysis and generating the report.
6. The name and version of the compliance software tool.

R405.4.3 Additional documentation. The *code official* shall be permitted to require the following documents:

1. Documentation of the building component characteristics of the *standard reference design.*
2. A certification signed by the builder providing the building component characteristics of the *proposed design* as given in Table R405.5.2(1).
3. Documentation of the actual values used in the software calculations for the *proposed design.*

R405.5 Calculation procedure. Calculations of the performance design shall be in accordance with Sections R405.5.1 and R405.5.2.

R405.5.1 General. Except as specified by this section, the *standard reference design* and *proposed design* shall be configured and analyzed using identical methods and techniques.

R405.5.2 Residence specifications. The *standard reference design* and *proposed design* shall be configured and analyzed as specified by Table R405.5.2(1). Table R405.5.2(1) shall include, by reference, all notes contained in Table R402.1.2.

TABLE R405.5.2(1)
SPECIFICATIONS FOR THE STANDARD REFERENCE AND PROPOSED DESIGNS

BUILDING COMPONENT	STANDARD REFERENCE DESIGN	PROPOSED DESIGN
Above-grade walls	Type: mass, where the proposed wall is a mass wall; otherwise, wood frame.	As proposed
	Gross area: same as proposed.	As proposed
	U-factor: as specified in Table R402.1.4.	As proposed
	Solar absorptance = 0.75.	As proposed
	Emittance = 0.90.	As proposed
Basement and crawl space walls	Type: same as proposed.	As proposed
	Gross area: same as proposed.	As proposed
	U-factor: as specified in Table R402.1.4, with the insulation layer on the interior side of the walls.	As proposed
Above-grade floors	Type: wood frame.	As proposed
	Gross area: same as proposed.	As proposed
	U-factor: as specified in Table R402.1.4.	As proposed
Ceilings	Type: wood frame.	As proposed
	Gross area: same as proposed.	As proposed
	U-factor: as specified in Table R402.1.4.	As proposed
Roofs	Type: composition shingle on wood sheathing.	As proposed
	Gross area: same as proposed.	As proposed
	Solar absorptance = 0.75.	As proposed
	Emittance = 0.90.	As proposed
Attics	Type: vented with an aperture of 1 ft^2 per 300 ft^2 of ceiling area.	As proposed
Foundations	Type: same as proposed.	As proposed
	Foundation wall area above and below grade and soil characteristics: same as proposed.	As proposed
Opaque doors	Area: 40 ft^2.	As proposed
	Orientation: North.	As proposed
	U-factor: same as fenestration as specified Table R402.1.4.	As proposed
Vertical fenestration other than opaque doors	Total area[h] = (a) The proposed glazing area, where the proposed glazing area is less than 15 percent of the conditioned floor area (b) 15 percent of the conditioned floor area, where the proposed glazing area is 15 percent or more of the conditioned floor area.	As proposed
	Orientation: equally distributed to four cardinal compass orientations (N, E, S & W).	As proposed
	U-factor: as specified in Table R402.1.4.	As proposed
	SHGC: as specified in Table R402.1.2 except for climate zones without an SHGC requirement, the SHGC shall be equal to 0.40.	As proposed
	Interior shade fraction: 0.92-(0.21 × SHGC for the standard reference design).	Interior shade fraction: 0.92-(0.21 × SHGC as proposed)
	External shading: none.	As proposed
Skylights	None.	As proposed
Thermally isolated sunrooms	None.	As proposed
Air exchange rate	The air leakage rate at a pressure of 0.2 inch w.g. (50 Pa) shall be *Climate Zones* 1 and 2: 5 air changes per hour. *Climate Zones* 3 through 8: 3 air changes per hour. The mechanical ventilation rate shall be in addition to the air leakage rate and shall be the same as in the proposed design, but not greater than $0.01 \times CFA + 7.5 \times (N_{br} + 1)$ where: *CFA* = conditioned floor area, ft^2. N_{br} = number of bedrooms. Energy recovery shall not be assumed for mechanical ventilation.	The measured air exchange rate.[a] The mechanical ventilation rate[b] shall be in addition to the air leakage rate and shall be as proposed.

(continued)

TABLE R405.5.2(1)—continued
SPECIFICATIONS FOR THE STANDARD REFERENCE AND PROPOSED DESIGNS

BUILDING COMPONENT	STANDARD REFERENCE DESIGN	PROPOSED DESIGN
Mechanical ventilation	Where mechanical ventilation is not specified in the proposed design: None Where mechanical ventilation is specified in the proposed design, the annual vent fan energy use, in units of kWh/yr, shall equal: $(1/e_f) \times [0.0876 \times CFA + 65.7 \times (N_{br} + 1)]$ where: e_f = the minimum exhaust fan efficacy, as specified in Table R403.6.1, corresponding to a flow rate of $0.01 \times CFA + 7.5 \times (N_{br}+1)$ CFA = conditioned floor area, ft^2. N_{br} = number of bedrooms.	As proposed
Internal gains	IGain, in units of Btu/day per dwelling unit, shall equal: $17{,}900 + 23.8 \times CFA + 4{,}104 \times N_{br}$ where: CFA = conditioned floor area, ft^2. N_{br} = number of bedrooms.	Same as standard reference design.
Internal mass	Internal mass for furniture and contents: 8 pounds per square foot of floor area.	Same as standard reference design, plus any additional mass specifically designed as a thermal storage element[c] but not integral to the building envelope or structure.
Structural mass	For masonry floor slabs: 80 percent of floor area covered by R-2 carpet and pad, and 20 percent of floor directly exposed to room air.	As proposed
	For masonry basement walls: as proposed, but with insulation as specified in Table R402.1.4, located on the interior side of the walls.	As proposed
	For other walls, ceilings, floors, and interior walls: wood frame construction.	As proposed
Heating systems[d, e]	For other than electric heating without a heat pump: as proposed. Where the proposed design utilizes electric heating without a heat pump, the standard reference design shall be an air source heat pump meeting the requirements of Section C403 of the IECC—Commercial Provisions. Capacity: sized in accordance with Section R403.7.	As proposed
Cooling systems[d, f]	As proposed. Capacity: sized in accordance with Section R403.7.	As proposed
Service water heating[d, e, f, g]	As proposed. Use: same as proposed design.	As proposed Use, in units of gal/day = $30 + (10 \times N_{br})$ where: N_{br} = number of bedrooms.
Thermal distribution systems	Duct insulation: in accordance with Section R403.3.1. A thermal distribution system efficiency (DSE) of 0.88 shall be applied to both the heating and cooling system efficiencies for all systems other than tested duct systems. **Exception:** For nonducted heating and cooling systems that do not have a fan, the standard reference design thermal distribution system efficiency (DSE) shall be 1. For tested duct systems, the leakage rate shall be 4 cfm (113.3 L/min) per 100 ft^2 (9.29 m^2) of *conditioned floor area* at a pressure of differential of 0.1 inch w.g. (25 Pa).	Duct insulation: as proposed. As tested or, where not tested, as specified in Table R405.5.2(2)
Thermostat	Type: Manual, cooling temperature setpoint = 75°F; heating temperature setpoint = 72°F.	Same as standard reference design.

(continued)

TABLE R405.5.2(1)—continued
SPECIFICATIONS FOR THE STANDARD REFERENCE AND PROPOSED DESIGNS

For SI: 1 square foot = 0.93 m^2, 1 British thermal unit = 1055 J, 1 pound per square foot = 4.88 kg/m^2, 1 gallon (US) = 3.785 L, °C = (°F - 32)/1.8, 1 degree = 0.79 rad.

a. Where required by the code official, testing shall be conducted by an approved party. Hourly calculations as specified in the ASHRAE *Handbook of Fundamentals*, or the equivalent, shall be used to determine the energy loads resulting from infiltration.

b. The combined air exchange rate for infiltration and mechanical ventilation shall be determined in accordance with Equation 43 of 2001 ASHRAE *Handbook of Fundamentals*, page 26.24 and the "Whole-house Ventilation" provisions of 2001 ASHRAE *Handbook of Fundamentals*, page 26.19 for intermittent mechanical ventilation.

c. Thermal storage element shall mean a component that is not part of the floors, walls or ceilings that is part of a passive solar system, and that provides thermal storage such as enclosed water columns, rock beds, or phase-change containers. A thermal storage element shall be in the same room as fenestration that faces within 15 degrees (0.26 rad) of true south, or shall be connected to such a room with pipes or ducts that allow the element to be actively charged.

d. For a proposed design with multiple heating, cooling or water heating systems using different fuel types, the applicable standard reference design system capacities and fuel types shall be weighted in accordance with their respective loads as calculated by accepted engineering practice for each equipment and fuel type present.

e. For a proposed design without a proposed heating system, a heating system having the prevailing federal minimum efficiency shall be assumed for both the standard reference design and proposed design.

f. For a proposed design home without a proposed cooling system, an electric air conditioner having the prevailing federal minimum efficiency shall be assumed for both the standard reference design and the proposed design.

g. For a proposed design with a nonstorage-type water heater, a 40-gallon storage-type water heater having the prevailing federal minimum energy factor for the same fuel as the predominant heating fuel type shall be assumed. For a proposed design without a proposed water heater, a 40-gallon storage-type water heater having the prevailing federal minimum efficiency for the same fuel as the predominant heating fuel type shall be assumed for both the proposed design and standard reference design.

h. For residences with conditioned basements, R-2 and R-4 residences, and for townhouses, the following formula shall be used to determine glazing area:

$AF = A_s \times FA \times F$

where:

AF = Total glazing area.

A_s = Standard reference design total glazing area.

FA = (Above-grade thermal boundary gross wall area)/(above-grade boundary wall area + 0.5 × below-grade boundary wall area).

F = (above-grade thermal boundary wall area)/(above-grade thermal boundary wall area + common wall area) or 0.56, whichever is greater.

and where:

Thermal boundary wall is any wall that separates conditioned space from unconditioned space or ambient conditions.

Above-grade thermal boundary wall is any thermal boundary wall component not in contact with soil.

Below-grade boundary wall is any thermal boundary wall in soil contact.

Common wall area is the area of walls shared with an adjoining dwelling unit. L and CFA are in the same units.

TABLE R405.5.2(2)
DEFAULT DISTRIBUTION SYSTEM EFFICIENCIES FOR PROPOSED DESIGNS[a]

DISTRIBUTION SYSTEM CONFIGURATION AND CONDITION	FORCED AIR SYSTEMS	HYDRONIC SYSTEMS[b]
Distribution system components located in unconditioned space	—	0.95
Untested distribution systems entirely located in conditioned space[c]	0.88	1
"Ductless" systems[d]	1	—

For SI: 1 cubic foot per minute = 0.47 L/s, 1 square foot = 0.093 m^2, 1 pound per square inch = 6895 Pa, 1 inch water gauge = 1250 Pa.

a. Default values in this table are for untested distribution systems, which must still meet minimum requirements for duct system insulation.

b. Hydronic systems shall mean those systems that distribute heating and cooling energy directly to individual spaces using liquids pumped through closed-loop piping and that do not depend on ducted, forced airflow to maintain space temperatures.

c. Entire system in conditioned space shall mean that no component of the distribution system, including the air-handler unit, is located outside of the conditioned space.

d. Ductless systems shall be allowed to have forced airflow across a coil but shall not have any ducted airflow external to the manufacturer's air-handler enclosure.

R405.6 Calculation software tools. Calculation software, where used, shall be in accordance with Sections R405.6.1 through R405.6.3.

R405.6.1 Minimum capabilities. Calculation procedures used to comply with this section shall be software tools capable of calculating the annual energy consumption of all building elements that differ between the *standard reference design* and the *proposed design* and shall include the following capabilities:

1. Computer generation of the *standard reference design* using only the input for the *proposed design.* The calculation procedure shall not allow the user to directly modify the building component characteristics of the *standard reference design.*
2. Calculation of whole-building (as a single *zone*) sizing for the heating and cooling equipment in the *standard reference design* residence in accordance with Section R403.6.7.
3. Calculations that account for the effects of indoor and outdoor temperatures and part-load ratios on the performance of heating, ventilating and air-conditioning equipment based on climate and equipment sizing.
4. Printed *code official* inspection checklist listing each of the *proposed design* component characteristics from Table R405.5.2(1) determined by the analysis to provide compliance, along with their respective performance ratings such as *R*-value, *U*-factor, SHGC, HSPF, AFUE, SEER and EF.

R405.6.2 Specific approval. Performance analysis tools meeting the applicable provisions of Section R405 shall be permitted to be *approved.* Tools are permitted to be *approved* based on meeting a specified threshold for a jurisdiction. The *code official* shall be permitted to approve such tools for a specified application or limited scope.

R405.6.3 Input values. When calculations require input values not specified by Sections R402, R403, R404 and R405, those input values shall be taken from an *approved* source.

SECTION R406
ENERGY RATING INDEX COMPLIANCE ALTERNATIVE

R406.1 Scope. This section establishes criteria for compliance using an Energy Rating Index (ERI) analysis.

R406.2 Mandatory requirements. Compliance with this section requires that the provisions identified in Sections R401 through R404 indicated as "Mandatory" and Section R403.5.3 be met. The *building thermal envelope* shall be greater than or equal to levels of efficiency and *Solar Heat Gain Coefficients* in Table 402.1.1 or 402.1.3 of the 2009 *International Energy Conservation Code.*

Exception: Supply and return ducts not completely inside the *building thermal envelope* shall be insulated to an *R*-value of not less than R-6.

R406.3 Energy Rating Index. The Energy Rating Index (ERI) shall be determined in accordance with RESNET/ICC 301 except for buildings covered by the *International Residential Code*, the ERI Reference Design Ventilation rate shall be in accordance with Equation 4-1.

Ventilation rate, CFM = (0.01 × total square foot area of house) + [7.5 × (number of bedrooms + 1)]

(Equation 4-1)

Energy used to recharge or refuel a vehicle used for transportation on roads that are not on the building site shall not be included in the *ERI reference design* or the *rated design.*

R406.4 ERI-based compliance. Compliance based on an ERI analysis requires that the *rated design* be shown to have an ERI less than or equal to the appropriate value indicated in Table R406.4 when compared to the *ERI reference design.*

TABLE R406.4
MAXIMUM ENERGY RATING INDEX

CLIMATE ZONE	ENERGY RATING INDEX[a]
1	57
2	57
3	57
4	62
5	61
6	61
7	58
8	58

a. Where on-site renewable energy is included for compliance using the ERI analysis of Section R406.4, the building shall meet the mandatory requirements of Section R406.2, and the building thermal envelope shall be greater than or equal to the levels of efficiency and SHGC in Table R402.1.2 or Table R402.1.4 of the 2015 *International Energy Conservation Code.*

R406.5 Verification by approved agency. Verification of compliance with Section R406 shall be completed by an *approved* third party.

R406.6 Documentation. Documentation of the software used to determine the ERI and the parameters for the *residential building* shall be in accordance with Sections R406.6.1 through R406.6.3.

R406.6.1 Compliance software tools. Software tools used for determining ERI shall be Approved Software Rating Tools in accordance with RESNET/ICC 301.

R406.6.2 Compliance report. Compliance software tools shall generate a report that documents that the ERI of the *rated design* complies with Sections R406.3 and R406.4. The compliance documentation shall include the following information:

1. Address or other identification of the residential building.
2. An inspection checklist documenting the building component characteristics of the *rated design.* The inspection checklist shall show results for both the *ERI reference design* and the *rated design*, and shall document all inputs entered by the user necessary to reproduce the results.

3. Name of individual completing the compliance report.
4. Name and version of the compliance software tool.

Exception: Where an otherwise identical building model is offered in multiple orientations, compliance for any orientation shall be permitted by documenting that the building meets the performance requirements in each of the four (north, east, south and west) cardinal orientations.

R406.6.3 Additional documentation. The *code official* shall be permitted to require the following documents:

1. Documentation of the building component characteristics of the *ERI reference design*.
2. A certification signed by the builder providing the building component characteristics of the *rated design*.
3. Documentation of the actual values used in the software calculations for the *rated design*.

R406.6.4 Specific approval. Performance analysis tools meeting the applicable sections of Section R406 shall be *approved*. Documentation demonstrating the approval of performance analysis tools in accordance with Section R406.6.1 shall be provided.

R406.6.5 Input values. Where calculations require input values not specified by Sections R402, R403, R404 and R405, those input values shall be taken from RESNET/ ICC 301.

CHAPTER 5 [RE]

EXISTING BUILDINGS

User note:

About this chapter: *Many buildings are renovated or altered in numerous ways that could affect the energy use of the building as a whole. Chapter 5 requires the application of certain parts of Chapter 4 in order to maintain, if not improve, the conservation of energy by the renovated or altered building.*

SECTION R501 GENERAL

R501.1 Scope. The provisions of this chapter shall control the *alteration*, *repair*, *addition* and change of occupancy of existing *buildings* and structures.

R501.1.1 Additions, alterations, or repairs: General. *Additions*, *alterations*, or *repairs* to an existing *building*, *building* system or portion thereof shall comply with Section R502, R503 or R504. Unaltered portions of the existing *building* or *building* supply system shall not be required to comply with this code.

R501.2 Existing buildings. Except as specified in this chapter, this code shall not be used to require the removal, *alteration* or abandonment of, nor prevent the continued use and maintenance of, an existing *building* or *building* system lawfully in existence at the time of adoption of this code.

R501.3 Maintenance. *Buildings* and structures, and parts thereof, shall be maintained in a safe and sanitary condition. Devices and systems that are required by this code shall be maintained in conformance to the code edition under which installed. The owner or the owner's authorized agent shall be responsible for the maintenance of *buildings* and structures. The requirements of this chapter shall not provide the basis for removal or abrogation of energy conservation, fire protection and safety systems and devices in existing structures.

R501.4 Compliance. *Alterations*, *repairs*, *additions* and changes of occupancy to, or relocation of, existing *buildings* and structures shall comply with the provisions for *alterations*, *repairs*, *additions* and changes of occupancy or relocation, respectively, in this code and the *International Residential Code*, *International Building Code*, *International Existing Building Code*, *International Fire Code*, *International Fuel Gas Code*, *International Mechanical Code*, *International Plumbing Code*, *International Property Maintenance Code*, *International Private Sewage Disposal Code* and NFPA 70.

R501.5 New and replacement materials. Except as otherwise required or permitted by this code, materials permitted by the applicable code for new construction shall be used. Like materials shall be permitted for *repairs*, provided that hazards to life, health or property are not created. Hazardous materials shall not be used where the code for new construction would not allow their use in *buildings* of similar occupancy, purpose and location.

R501.6 Historic buildings. Provisions of this code relating to the construction, *repair*, *alteration*, restoration and movement of structures, and *change of occupancy* shall not be mandatory for *historic buildings* provided that a report has been submitted to the *code official* and signed by the owner, a *registered design professional*, or a representative of the State Historic Preservation Office or the historic preservation authority having jurisdiction, demonstrating that compliance with that provision would threaten, degrade or destroy the historic form, fabric or function of the *building*.

SECTION R502 ADDITIONS

R502.1 General. *Additions* to an existing *building*, *building* system or portion thereof shall conform to the provisions of this code as those provisions relate to new construction without requiring the unaltered portion of the existing *building* or *building* system to comply with this code. *Additions* shall not create an unsafe or hazardous condition or overload existing *building* systems. An *addition* shall be deemed to comply with this code where the *addition* alone complies, where the existing *building* and *addition* comply with this code as a single *building*, or where the *building* with the *addition* does not use more energy than the existing *building*. *Additions* shall be in accordance with Section R502.1.1 or R502.1.2.

R502.1.1 Prescriptive compliance. *Additions* shall comply with Sections R502.1.1.1 through R502.1.1.4.

R502.1.1.1 Building envelope. New *building* envelope assemblies that are part of the *addition* shall comply with Sections R402.1, R402.2, R402.3.1 through R402.3.5, and R402.4.

Exception: Where *unconditioned space* is changed to *conditioned space*, the *building* envelope of the addition shall comply where the Total UA, as determined in Section R402.1.5, of the existing *building* and the *addition*, and any *alterations* that are part of the project, is less than or equal to the Total UA generated for the existing *building*.

R502.1.1.2 Heating and cooling systems. New heating, cooling and duct systems that are part of the *addition* shall comply with Section R403.

Exception: Where ducts from an existing heating and cooling system are extended to an *addition*, duct systems with less than 40 linear feet (12.19 m) in *unconditioned spaces* shall not be required to be tested in accordance with Section R403.3.3.

R502.1.1.3 Service hot water systems. New service hot water systems that are part of the *addition* shall comply with Section R403.5.

R502.1.1.4 Lighting. New lighting systems that are part of the *addition* shall comply with Section R404.1.

R502.1.2 Existing plus addition compliance (Simulated Performance Alternative). Where *unconditioned space* is changed to *conditioned space*, the *addition* shall comply where the annual energy cost or energy use of the *addition* and the existing *building*, and any *alterations* that are part of the project, is less than or equal to the annual energy cost of the existing *building* when modeled in accordance with Section R405. The *addition* and any *alterations* that are part of the project shall comply with Section R405 in its entirety.

SECTION R503
ALTERATIONS

R503.1 General. *Alterations* to any *building* or structure shall comply with the requirements of the code for new construction. *Alterations* shall be such that the existing *building* or structure is not less conforming to the provisions of this code than the existing *building* or structure was prior to the *alteration*.

Alterations to an existing *building*, *building* system or portion thereof shall conform to the provisions of this code as they relate to new construction without requiring the unaltered portions of the existing *building* or *building* system to comply with this code. *Alterations* shall not create an unsafe or hazardous condition or overload *existing* building systems. *Alterations* shall be such that the existing *building* or structure does not use more energy than the existing *building* or structure prior to the *alteration*. *Alterations* to existing *buildings* shall comply with Sections R503.1.1 through R503.2.

R503.1.1 Building envelope. *Building* envelope assemblies that are part of the *alteration* shall comply with Section R402.1.2 or R402.1.4, Sections R402.2.1 through R402.2.13, R402.3.1, R402.3.2, R402.4.3 and R402.4.5.

Exception: The following *alterations* shall not be required to comply with the requirements for new construction provided that the energy use of the *building* is not increased:

1. Storm windows installed over existing fenestration.
2. Existing ceiling, wall or floor cavities exposed during construction provided that these cavities are filled with insulation.
3. Construction where the existing roof, wall or floor cavity is not exposed.
4. Roof re-cover.
5. Roofs without insulation in the cavity and where the sheathing or insulation is exposed during reroofing shall be insulated either above or below the sheathing.
6. Surface-applied window film installed on existing single pane fenestration assemblies to reduce solar heat gain provided that the code does not require the glazing or fenestration assembly to be replaced.

R503.1.1.1 Replacement fenestration. Where some or all of an existing fenestration unit is replaced with a new fenestration product, including sash and glazing, the replacement fenestration unit shall meet the applicable requirements for *U*-factor and SHGC as specified in Table R402.1.2. Where more than one replacement *fenestration* unit is to be installed, an area-weighted average of the *U*-factor, SHGC or both of all replacement *fenestration* units shall be an alternative that can be used to show compliance.

R503.1.2 Heating and cooling systems. New heating, cooling and duct systems that are part of the *alteration* shall comply with Section R403.

Exception: Where ducts from an existing heating and cooling system are extended, duct systems with less than 40 linear feet (12.19 m) in *unconditioned spaces* shall not be required to be tested in accordance with Section R403.3.3.

R503.1.3 Service hot water systems. New service hot water systems that are part of the *alteration* shall comply with Section R403.5.

R503.1.4 Lighting. New lighting systems that are part of the *alteration* shall comply with Section R404.1.

Exception: *Alterations* that replace less than 50 percent of the luminaires in a space, provided that such *alterations* do not increase the installed interior lighting power.

R503.2 Change in space conditioning. Any nonconditioned or low-energy space that is altered to become *conditioned space* shall be required to be brought into full compliance with this code.

Exception: Where the simulated performance option in Section R405 is used to comply with this section, the annual energy cost of the *proposed design* is permitted to be 110 percent of the annual energy cost otherwise allowed by Section R405.3.

SECTION R504
REPAIRS

R504.1 General. *Buildings*, structures and parts thereof shall be repaired in compliance with Section R501.3 and this section. Work on nondamaged components necessary for the required *repair* of damaged components shall be considered to be part of the *repair* and shall not be subject to the requirements for *alterations* in this chapter. Routine maintenance required by Section R501.3, ordinary *repairs* exempt from *permit*, and abatement of wear due to normal service conditions shall not be subject to the requirements for *repairs* in this section.

R504.2 Application. For the purposes of this code, the following shall be considered to be *repairs*:

1. Glass-only replacements in an existing sash and frame.
2. Roof *repairs*.
3. *Repairs* where only the bulb, ballast or both within the existing luminaires in a space are replaced provided that the replacement does not increase the installed interior lighting power.

SECTION R505
CHANGE OF OCCUPANCY OR USE

R505.1 General. Spaces undergoing a change in occupancy that would result in an increase in demand for either fossil fuel or electrical energy shall comply with this code.

R505.2 General. Any space that is converted to a dwelling unit or portion thereof from another use or occupancy shall comply with this code.

Exception: Where the simulated performance option in Section R405 is used to comply with this section, the annual energy cost of the *proposed design* is permitted to be 110 percent of the annual energy cost allowed by Section R405.3.

CHAPTER 6 [RE]
REFERENCED STANDARDS

User note:

About this chapter: *This code contains numerous references to standards promulgated by other organizations that are used to provide requirements for materials and methods of construction. Chapter 6 contains a comprehensive list of all standards that are referenced in this code. These standards, in essence, are part of this code to the extent of the reference to the standard.*

This chapter lists the standards that are referenced in various sections of this document. The standards are listed herein by the promulgating agency of the standard, the standard identification, the effective date and title, and the section or sections of this document that reference the standard. The application of the referenced standards shall be as specified in Section R107.

AAMA

American Architectural Manufacturers Association
1827 Walden Office Square
Suite 550
Schaumburg, IL 60173-4268

AAMA/WDMA/CSA 101/I.S.2/A C440—17: North American Fenestration Standard/Specifications for Windows, Doors and Unit Skylights
R402.4.3

ACCA

Air Conditioning Contractors of America
2800 Shirlington Road, Suite 300
Arlington, VA 22206

Manual J—16: Residential Load Calculation Eighth Edition
R403.7

Manual S—14: Residential Equipment Selection
R403.7

APSP

The Association of Pool & Spa Professionals
2111 Eisenhower Avenue, Suite 500
Alexandria, VA 22314

ANSI/APSP/ICC 14—2014: American National Standard for Portable Electric Spa Energy Efficiency
R403.11

ANSI/APSP/ICC 15a—2011: American National Standard for Residential Swimming Pool and Spa Energy Efficiency—includes Addenda A Approved January 9, 2013
R403.12

ASHRAE

ASHRAE
1791 Tullie Circle NE
Atlanta, GA 30329

ASHRAE—2017: ASHRAE Handbook of Fundamentals
R402.1.5

ASHRAE—2001: 2001 ASHRAE Handbook of Fundamentals
Table R405.5.2(1)

ASHRAE 193—2010(RA 2014): Method of Test for Determining the Airtightness of HVAC Equipment
R403.3.2.1

ASTM

ASTM International
100 Barr Harbor Drive, P.O. Box C700
West Conshohocken, PA 19428-2959

C1363—11: Standard Test Method for Thermal Performance of Building Materials and Envelope Assemblies by Means of a Hot Box Apparatus
R303.1.4.1

ASTM—continued

E283—04(2012): Test Method for Determining the Rate of Air Leakage Through Exterior Windows, Curtain Walls and Doors Under Specified Pressure Differences Across the Specimen
R402.4.4

E779—10: Standard Test Method for Determining Air Leakage Rate by Fan Pressurization
R402.4.1.2

E1827—11: Standard Test Methods for Determining Airtightness of Building Using an Orifice Blower Door
R402.4.1.2

CSA

CSA Group
8501 East Pleasant Valley Road
Cleveland, OH 44131-5516

AAMA/WDMA/CSA 101/I.S.2/A440—17: North American Fenestration Standard/Specification for Windows, Doors and Unit Skylights
R402.4.3

CSA B55.1—2015: Test Method for Measuring Efficiency and Pressure Loss of Drain Water Heat Recovery Units
R403.5.4

CSA B55.2—2015: Drain Water Heat Recovery Units
R403.5.4

DASMA

Door & Access Systems Manufacturers Association
1300 Sumner Avenue
Cleveland, OH 44115-2851

105—2016: Test Method for Thermal Transmittance and Air Infiltration of Garage Doors and Rolling Doors
R303.1.3

HVI

Home Ventilating Institute
1000 North Rand Road, Suite 214
Wauconda, IL 60084

916—09: Airflow Test Procedure
Table R403.6.1

ICC

International Code Council, Inc.
500 New Jersey Avenue NW
6th Floor
Washington, DC 20001

ANSI/APSP/ICC 14—2014: American National Standard for Portable Electric Spa Energy Efficiency
R403.11

ANSI/APSP/ICC 15a—2011: American National Standard for Residential Swimming Pool and Spa Energy Efficiency—includes Addenda A Approved January 9, 2013
R403.12

IBC—18: International Building Code®
R201.3, R303.1.1, R303.2, R402.1.1, R501.4

ICC 400—17: Standard on the Design and Construction of Log Structures
R402.1, Table R402.5.1.1

IEBC—18: International Existing Building Code®
R501.4

IECC—18: International Energy Conservation Code®
R101.4.1, R403.8

IECC—15: 2015 International Energy Conservation Code®
Table R406.4

IECC—09: 2009 International Energy Conservation Code®
R406.2

ICC—continued

IECC—06: 2006 International Energy Conservation Code®
R202

IFC—18: International Fire Code®
R201.3, R501.4

IFGC—18: International Fuel Gas Code®
R201.3, R501.4

IMC—18: International Mechanical Code®
R201.3, R403.3.2, R403.3.6, R403.6, R501.4

IPC—18: International Plumbing Code®
R201.3, R501.4

IPSDC—18: International Private Sewage Disposal Code®
R501.4

IPMC—18: International Property Maintenance Code®
R501.4

IRC—18: International Residential Code®
R201.3, R303.1.1, R303.2, R402.1.1, R402.2.11, R403.3.2, R403.3.6, R403.6, R501.4

ANSI/RESNET/ICC 301—2014: Standard for the Calculation and Labeling of the Energy Performance of Low-rise Residential Buildings using an Energy Rating Index First Published March 7, 2014—Republished January 2016
R406.3

ANSI/RESNET/ICC 380—2016: Standard for Testing Airtightness for Building Enclosures, Airtightness of Heating and Cooling Air Distribution Systems and Airflow of Mechanical Ventilation Systems—Republished January 2016
R402.4.1.2

IEEE

Institute of Electrical and Electronic Engineers, Inc.
3 Park Avenue, 17th Floor
New York, NY 10016-5997

515.1—2012: IEEE Standard for the Testing, Design, Installation and Maintenance of Electrical Resistance Trace Heating for Commercial Applications
R403.5.1.2

NFPA

National Fire Protection Association
1 Batterymarch Park
Quincy, MA 02169-7471

70—17: National Electrical Code
R501.4

NFRC

National Fenestration Rating Council, Inc.
6305 Ivy Lane, Suite 140
Greenbelt, MD 20770

100—2017: Procedure for Determining Fenestration Products *U*-factors
R303.1.3

200—2017: Procedure for Determining Fenestration Product Solar Heat Gain Coefficients and Visible Transmittance at Normal Incidence
R303.1.3

400—2017: Procedure for Determining Fenestration Product Air Leakage
R402.4.3

RESNET

Residential Energy Services Network, Inc.
P.O. Box 4561
Oceanside, CA 92052-4561

ANSI/RESNET/ICC 301—2014: Standard for the Calculation and Labeling of the Energy Performance of Low-rise Residential Buildings using an Energy Rating Index First Published March 7, 2014—Republished January 2016
R406.3, R406.6.1, R406.6.5

RESNET—continued

ANSI/RESNET/ICC 380—2016: Standard for Testing Airtightness for Building Enclosures, Airtightness of Heating and Cooling Air Distribution Systems, and Airflow of Mechanical Ventilation Systems—Republished January 2016

R402.4.1.2

UL

UL LLC
333 Pfingsten Road
Northbrook, IL 60062

127—11: Standard for Factory Built Fireplaces—with Revisions through May 2015

R402.4.2

515—11: Electrical Resistance Heat Tracing for Commercial and Industrial Applications Including Revisions through July 2015

R403.5.1.2

US-FTC

United States-Federal Trade Commission
600 Pennsylvania Avenue NW
Washington, DC 20580

CFR Title 16 (2015): R-value Rule

R303.1.4

WDMA

Window and Door Manufacturers Association
2025 M Street NW, Suite 800
Washington, DC 20036-3309

AAMA/WDMA/CSA 101/I.S.2/A440—17: North American Fenestration Standard/Specification for Windows, Doors and Unit Skylights

R402.4.3

APPENDIX RA

SOLAR-READY PROVISIONS—DETACHED ONE- AND TWO-FAMILY DWELLINGS AND TOWNHOUSES

The provisions contained in this appendix are not mandatory unless specifically referenced in the adopting ordinance.

User note:

About this appendix: *Harnessing the heat or radiation from the sun's rays is a method to reduce the energy consumption of a building. Although Appendix RA does not require solar systems to be installed for a building, it does require the space(s) for installing such systems, providing pathways for connections and requiring adequate structural capacity of roof systems to support the systems.*

SECTION RA101 SCOPE

RA101.1 General. These provisions shall be applicable for new construction where solar-ready provisions are required.

SECTION RA102 GENERAL DEFINITION

SOLAR-READY ZONE. A section or sections of the roof or building overhang designated and reserved for the future installation of a solar photovoltaic or solar thermal system.

SECTION RA103 SOLAR-READY ZONE

RA103.1 General. New detached one- and two-family dwellings, and townhouses with not less than 600 square feet (55.74 m^2) of roof area oriented between 110 degrees and 270 degrees of true north shall comply with Sections RA103.2 through RA103.8.

Exceptions:

1. New residential buildings with a permanently installed on-site renewable energy system.
2. A building with a solar-ready zone that is shaded for more than 70 percent of daylight hours annually.

RA103.2 Construction document requirements for solar-ready zone. Construction documents shall indicate the solar-ready zone.

RA103.3 Solar-ready zone area. The total solar-ready zone area shall be not less than 300 square feet (27.87 m^2) exclusive of mandatory access or set back areas as required by the *International Fire Code*. New townhouses three stories or less in height above grade plane and with a total floor area less than or equal to 2,000 square feet (185.8 m^2) per dwelling shall have a solar-ready zone area of not less than 150 square feet (13.94 m^2). The solar-ready zone shall be composed of areas not less than 5 feet (1524 mm) in width and not less than 80 square feet (7.44 m^2) exclusive of access or set back areas as required by the *International Fire Code*.

RA103.4 Obstructions. Solar-ready zones shall be free from obstructions, including but not limited to vents, chimneys, and roof-mounted equipment.

RA103.5 Roof load documentation. The structural design loads for roof dead load and roof live load shall be clearly indicated on the construction documents.

RA103.6 Interconnection pathway. Construction documents shall indicate pathways for routing of conduit or plumbing from the solar-ready zone to the electrical service panel or service hot water system.

RA103.7 Electrical service reserved space. The main electrical service panel shall have a reserved space to allow installation of a dual pole circuit breaker for future solar electric installation and shall be labeled "For Future Solar Electric." The reserved space shall be positioned at the opposite (load) end from the input feeder location or main circuit location.

RA103.8 Construction documentation certificate. A permanent certificate, indicating the solar-ready zone and other requirements of this section, shall be posted near the electrical distribution panel, water heater or other conspicuous location by the builder or registered design professional.

INDEX

E

F

G

H

EDITORIAL CHANGES – FIFTH PRINTING

Page C-37, **TABLE C402.3:** row 1 now reads . . .

Three-year-aged solar reflectance[b] of 0.55 and 3-year aged thermal emittance[c] of 0.75

Page R-42, **Section R405.6.1:** Item 2, line 4 now reads . . . with Section R403.6.7.

For the complete errata history of this code, please visit: https://www.iccsafe.org/errata-central/